X-Ray Vision: Nikola Tesla On Roentgen Rays

X-Ray Vision: Nikola Tesla On Roentgen Rays

by Nikola Tesla

Table of Contents

On Roentgen Rays

Electrical Review — March 11, 1896

One can not help looking at that little bulb of Crookes with a feeling akin to awe, when he considers all that it has done for scientific progress — first, the magnificent results obtained by its originator; next, the brilliant work of Lenard, and finally the wonderful achievements of Roentgen. Possibly it may still contain a grateful Asmodeus, who will be lot out of his narrow prison cell by a lucky student. At times it has seemed to me as though I myself heard a whispering voice, and I have searched eagerly among my dusty bulbs and bottles. I fear my imagination has deceived me, but there they are still, my dusty bulbs, and I am still listening hopefully.

After repeating Professor Roentgen's beautiful experiments, I have devoted my energies to the investigation of the nature of the radiations and to the perfecting of the means for their production. The following is a brief statement which, I hope, will be useful, of the methods employed and of the most notable results arrived at in these two directions.

In order to produce the most intense effects we have first to consider that, whatever their nature, they depend necessarily on the intensity of the cathode streams. These again being dependent on the magnitude of the potential, it follows that the highest attainable electrical pressure is desirable.

To obtain high potentials we may avail ourselves of an ordinary induction coil, or of a static machine, or of a disruptive discharge coil. I have the impression that most of the results in Europe have been arrived at through the employment of a static machine or Ruhmkorff coil. But since these appliances can produce only a comparatively small potential, we are naturally thrown on the use of the disruptive discharge coil as the most effective apparatus. With this .there is practically no limit to the spark length, and the only requirement is that the experimenter should possess a certain knowledge and skill in the adjustments of the circuits, particularly as to resonance, as I have pointed out in my earlier writings on this subject.

After constructing a disruptive coil suitable for any kind of current supply, direct or alternating the experimenter comes to the consideration as to what kind of bulb to employ. Clearly, if we put two electrodes in a bulb, or use one inside and another outside electrode, we limit the potential, for the presence not only of the anode but of any conducting object has the effect of reducing the practicable potential on the cathode. Thus, to secure the result aimed at, one is driven to the acceptance of a single electrode bulb; the other terminal being as far remote as possible.

Obviously, an inside electrode should be employed to get the highest velocity of the cathode streams, for the bulbs without inside terminals are much less efficient for this special object in consequence of the loss through the glass. A popular error seems to exist in regard to the concentration of the *rays* by concave electrodes. This, if anything, is a disadvantage. There are certain specific arrangements of the disruptive coil and circuits, condensers and static screens for the bulb, on which I have given full particulars on previous occasions.

Having selected the induction apparatus and type of bulb, tie next important consideration is the vacuum. On this subject I am able to make known a fact with which I have long been acquainted, and of which I have taken advantage in the production of vacuum jackets and all sorts of incandescent bulbs, and which I subsequently found to be of the utmost importance,. not to say essential, for the production of intense Roentgen shadows. I refer to a method of rarefaction by electrical means to any degree desirable far beyond that obtainable by mechanical appliances.

Though this result can be reached by the use of a static machine as, well as of an ordinary induction coil giving a sufficiently high potential, I have found that by far the most suitable apparatus, and one which secures the quickest action, is a disruptive coil: It is best to proceed in this way: The bulb is first exhausted by means of an ordinary vacuum pump to a rather high degree, though my experiences have shown that this is not absolutely necessary, as I have also found it possible to rarefy, beginning from low pressure. After being taken down from the pump, the bulb is attached to the terminal of the disruptive coil, preferably of high frequency of vibration, and usually the following phenomena are noted. First, there is a milky light spreading through the bulb, or possibly for a moment the glass becomes phosphorescent, if the bulb has been exhausted to a high degree. At any rate, the phosphorescence generally subsides quickly and the white light settles around .the electrode, whereupon a dark space forms at some distance from the latter. Shortly afterward the light assumes a reddish. color and the terminal grows very hot. This heating, however, is observed only with powerful apparatus. It is well to watch the bulb carefully and regulate the potential at this stage, as the electrode might be quickly consumed.

After some time the reddish light subsides, the streams becoming again white, whereupon they get weaker and weaker, wavering around the electrode until they finally disappear: Meanwhile, the phosphorescence of the glass grows more and more intense, and the spot where the stream strikes the wall becomes very hot, while the phosphorescence around the electrode ceases and the latter cools down to such an extent that the glass near it may be actually ice-cold to the touch. The gas in the bulb has then reached the required degree of rarefaction. The process may be hastened by repeated heating arid cooling and by the employment of a small electrode. It should be added that bulbs with external electrodes may be treated in the same way. It may be also of interest to state that under certain conditions, which I am investigating more closely, the pressure of the gas ,in a vessel may be augmented by electrical means.

I believe that the disintegration of the electrode, which invariably takes. place, is connected with a notable diminution of the temperature. From the point on, when the electrode gets cool, the bulb is in a very good condition for producing the Roentgen shadows. Whenever the electrode is equally, if not hotter than the glass, it is a sure indication that the vacuum is not high enough, or else that the electrode is too small. For very effective working, the inside surface of the wall, where the cathode stream strikes, should appear as if the glass were in a fluid state.

As a cooling medium I have found best to employ jets of cold air. By this means it is possible to operate successfully a bulb with a very thin wall, while the passage of the rays is not materially impeded.

I may state here that the experimenter need not be deterred from using a glass bulb, as I believe the opacity of glass as well as the transparency of aluminum are somewhat exaggerated, inasmuch as I have found that a very thin aluminum sheet throws a marked shadow, while, on the other hand, I have obtained impressions through a thick glass plate.

The above method is valuable not only as a means of obtaining the high vacua desired, but it is still more important, because the phenomena observed throw a light on the results obtained by Lenard and Roentgen.

Though the phenomenon of rarefaction under above conditions admits of different interpretations, the chief interest renters on one of them, to which I adhere — that is, on the actual expulsion of the particles through the walls of the bulb. I have lately observed that the latter commences to act properly upon the sensitive plate only from the point when the exhaustion begins to be noticeable, and the effects produced are the strongest when the process of exhaustion is most rapid, even though the phosphorescence might not appear particularly bright. Evidently, then, the two effects are closely connected, and I am getting more and more convinced that we have to deal with a stream of material particles, which strike the sensitive plate with great velocities. Taking as a basis the estimate of .Lord Kelvin on the speed of projected 'particles in a Crookes' bulb, we arrive easily by the employment of very high potentials to speeds of as much as a hundred kilometers a second. Now, again, the old question arises: Are the particles from the electrode ,or from the charged surface generally, including the case of an external electrode, projected through the glass or aluminum walls, or do they merely hit the .inner surface and cause particles from the outside of the wall to fly off, acting in a purely mechanical way, as when a row of ivory balls is struck? So far, most of the phenomena indicate that they are projected through the wall of the bulb, of whatever material it may be, and I am seeking for still more conclusive evidence in this direction.

It may not be known that even an ordinary streamer, breaking out suddenly and under great pressure from the terminal of a disruptive coil, passes through a thick glass plate as though the latter were not present. Unquestionably, with such coils pressures are practicable which will, project the particles in straight lines even under atmospheric pressure. I have obtained distinct impressions in free air, not by streamers, as some experimenters have done, using static machines or induction coils, but by actual projection, the formation of streamers being absolutely prevented by careful static screening.

A peculiar thing about the Roentgen rays is that from low frequency to the highest obtainable there seems to be no difference in the quality of the effects produced, except that they are more intense when the frequency is higher, which is very likely due to the fact that in such case the maximum pressures on the cathode are likewise higher. This is only possible on the assumption that the effects on the sensitive plate are due to projected particles, or else to vibrations far beyond any frequency which we are able to obtain by means of condenser discharges. A powerfully excited bulb is enveloped in a cloud of violet light, extending for more than a foot around it, but outside of this visible phenomenon there is no positive evidence of the existence of waves similar to those of light. On the other hand, the fact that the opacity bears some proportion to the density of the substance speaks strongly for material streams, and the same may be said of the effect discovered by Prof. J. J. Thomson. It is to be hoped that all doubts will shortly be dispelled.

A valuable evidence of the nature of the radiations and progress in the direction of obtaining strong impressions on the plate might be arrived at by perfecting plates especially sensitive to mechanical shock or impact. There are chemicals suitable for this, and the development in this direction may lead to the abandonment of the present plate. Furthermore, if we have to deal with stream. of material particles, it seems not impossible to project upon the plate a suitable substance to insure the best chemical action.

With apparatus as I have described, remarkable impressions on the plate are produced. An idea of the intensity of the effects may be gained when I mention that it is easy to obtain shadows with comparatively short exposures at distances of many feet, while at small distances and with thin objects, exposures of a few seconds are practicable. The annexed print is a shadow of a copper wire projected at a distance of 11 feet through a wooden cover over the sensitive plate. This was the first shadow taken with my improved apparatus in my laboratory. A similar impression was obtained through the body of the experimenter, a plate of glass; nearly three-sixteenths of an inch thick, a thickness of wood of fully two inches and through a distance of about four feet. I may remark, however, that when these impressions were taken, my apparatus was working under extremely unfavorable conditions, which admitted of so great improvements that I am hopeful to magnify the effects many times.

The bony structure of birds, rabbits and the like is shown within the least detail, and even the hollow of the bones is clearly visible. In a plate of a rabbit under exposure of an hour, not only every detail of the skeleton is visible, but likewise a clear outline of the abdominal cavity and the location of the lungs, the fur and many other features. Prints of even large birds show the feathers quite distinctly.

Clear shadows of the bones of human limbs are obtained by exposures ranging from a quarter of an hour to an hour, and some plates have shown such an amount of detail that it is almost impossible to believe that we have to deal with shadows only. For instance, a picture of a foot with a shoe on it was taken, and every fold of the leather, trousers, stocking, etc., is visible, while the flesh and bones stand out sharply. Through the body of the experimenter the shadows of small buttons and like objects are quickly obtained, while with an exposure of from one to one and a half hour the ribs, shoulder-bones and the bones of the upper arm appear dearly, as is shown in the annexed print. It is now demonstrated beyond any doubt that small metallic objects or bony or chalky deposits can be infallibly detected in any part of the body.

An outline of the skull is easily obtained with an exposure of 20 to 40 minutes. In one instance an exposure of 40 minutes gave dearly not only the outline, but the cavity of the eye, the chin and cheek and nasal bones, the lower jaw and connections to the upper one, the vertebral column and connections to the skull, the flesh and even the hair. By exposing the head to a powerful radiation strange effects have been noted. For instance, I find that there is a tendency to sleep and the time seems to pass away quickly. There is a general soothing effect, and I have: felt a sensation of warmth in the upper part of the head. An assistant independently confirmed the tendency to sleep and a quick lapse of time. Should these remarkable effects be verified by men with keener sense of observation, I shall still more firmly believe in the existence of material streams penetrating the skull. Thus it may be possible by these strange appliances to project a suitable chemical into any part of the body.

Roentgen advanced modestly his results, warning against too much hope. Fortunately his apprehensions were groundless, for, although we have to all appearance to deal with mere shadow projections, the possibilities of the application of his discovery are vast. I am happy to have contributed to the development of the great art he has created.

On Roentgen Rays the Latest Results

Electrical Review — March 18, 1896

To the Edition of Electrical Review

Permit me to say that I was slightly disappointed to note in your issue of March 11 the prominence you have deemed to accord to my youth and talent, while the ribs and other particulars of Fig. 1, which, with reference to the print accompanying my communication, I described as clearly visible, were kept modestly in the background: I also regretted to observe an error in one of the captions, the more so; as I must ascribe it to my own text. I namely stated on page 135, third column, seventh line: "A similar impression was obtained through the body of the experimenter, etc., through a distance of four feet." The impression here referred to was a similar one to that shown in Fig. 2, whereas the shadow in Fig. 1 was taken through a distance of 18 inches. I state this merely for the sake of correctness of my communication, but, as far as the general truth of the fact of taking such a shadow at the distance given is concerned, your caption might as well stand, for I am producing strong shadows at distances of 40 *feet.* I repeat, 40 feet and even more. Nor is ,this all. So strong are the actions on the film that provision must be made to guard the plates in my photographic a department, located on the floor above, a distance of fully :60 feet, from being spoiled by long exposure to the stray rays. Though during my investigations I have performed many experiments which seemed extraordinary, I am deeply astonished observing these unexpected manifestations, and still more so, as even now I see before me the possibility, not to say certitude, of augmenting the effects with my apparatus at least tenfold! What may we then expect? We have to deal here, evidently, with a radiation of astonishing power, and the inquiry into its nature becomes more and more interesting and important. Here is an unlooked-for result of an action which, though wonderful in itself, seemed feeble and entirely incapable of such expansion, and affords a good example of the fruitfulness of original discovery. These effects upon the sensitive plate at so great a distance I attribute to the employment of a bulb with a single terminal, which permits the use of practically any desired potential and the attainment of extraordinary speeds of the projected particles. With such a bulb it is also evident that the action upon a fluorescent screen is proportionately greater than when the usual kind of tube is employed, and I have already observed enough to feel sure that great developments are to be looked for in this direction. I consider Roentgen's discovery; of enabling us to see, by the use of a fluorescent screen, through an opaque substance, even a more beautiful one than the recording upon the plate.

Since my previous communication to you I have made considerable progress, and can presently announce one more result of importance. I have lately obtained shadows *by reflected rays only,* thus demonstrating beyond doubt that the Roentgen rays possess this property. One of the experiments may be cited here. A thick copper tube, about a foot long, was taken and one of its ends tightly closed by the plate-holder containing a sensitive plate, protected by a fiber cover as usual. Near the open end of the copper tube was placed a thick plate of glass at an angle of 45 degrees to the axis of the tube. A single-terminal bulb was then suspended above the glass plate at a distance of about eight inches, so that the bundle of rays fell upon the latter at in angle of 45 degrees, and the supposedly reflected rays passed along the axis of the copper tube. An exposure of 45 minutes gave a dear and sharp shadow of a metallic object. This shadow was produced by the reflected rays, as the direct action was absolutely excluded, it having been demonstrated that even under the severest tests with much stronger actions no impression whatever could be produced upon the film through a thickness of copper equal to that of the tube. Concluding from the intensity of the action by comparison with an equivalent effect due to the direct rays, I find that approximately two per cent of the latter were reflected from the glass plate in this experiment. I hope to be able to report shortly and more fully on this and other subjects.

In my attempts to contribute my humble share to the knowledge of the Roentgen phenomena, I am finding more and more evidence in support of the theory of moving material particles. It is not my intention, however, to advance at present any view as to the bearing of such a fact upon the present theory of light, but I merely seek to establish the fact of the existence of such material streams in so far as these isolated effects are concerned. I have already a great many indications of a bombardment occurring outside of the bulb, and I am arranging some crucial tests which, I hope, will be successful. The calculated velocities fully account for actions at distances of as much as 100 feet from the bulb, and that the projection through the glass takes place seems evident from the process of exhaustion, which I have described in my previous communication. An experiment which is illustrative in this respect, and which I intended to mention, is the following: If we attach a fairly exhausted bulb containing an electrode to the terminal of a disruptive coil, we observe small streamers breaking through the sides of the glass. Usually such a streamer will break through the seal and crack the bulb, whereupon the vacuum is impaired; but, if the seal is placed above the terminal, or if some other provision is made to prevent the streamer from passing through the glass at that point, it often occurs that the stream breaks out through the side of the bulb, producing a fine hole. Now, the extraordinary thing is that, in spite of the connection to the outer atmosphere, the air can not rush into the bulb as long as the hole is very small. The glass at the place where the rupture has occurred may grow very hot — to such a degree as to soften; but it will not collapse, but rather bulge out, showing that a pressure from the inside greater than that of the atmosphere exists. On frequent occasions I have observed that the glass bulges out and the hole, through which the streamer rushes out, becomes so large. as to be perfectly discernible to the eye. As the matter is expelled from the bulb the rarefaction increases and the streamer becomes less and less intense, whereupon the glass doses again, hermetically sealing the opening. The process of rarefaction, nevertheless, continues, streamers being still visible on the heated place until the highest degree of exhaustion is reached, whereupon they may disappear. Here, then, we have a positive evidence that matter is being expelled through the walls of the glass.

When working with highly strained bulbs I frequently experience a sudden, and sometimes even painful, shock in the eye. Such shocks may occur so often that the eye gets inflamed, and one can not be considered overcautious if he abstains from watching the bulb too closely. I see in these shocks a further evidence of larger particles being thrown off from the bulb.

On Reflected Roentgen Rays
Electrical Review — April 1, 1896

In previous communications in regard to the effects discovered by Roentgen, I have confined myself to giving barely a brief outline of the most noteworthy results arrived at in the course of my investigations. To state truthfully, I have ventured to express myself, the first time, after some hesitation and consequent delay, and only when I had gained the conviction that the information I had to convey was a needful one; for, in common with others, I was not quite able to free myself of a certain feeling which one must experience when he is trespassing on ground not belonging to him. The discoverer would naturally himself arrive at most of the facts in due time, and a courteous restraint in the announcement of the results on the part of his co-workers would not be amiss. How many have sinned against me by proclaiming their achievements just as I was good and ready to do it myself! But these discoveries of Roentgen, exactly of the order of the telescope and microscope, his seeing through a great thickness of an opaque substance, his recording on a sensitive plate of objects otherwise invisible, were so beautiful and fascinating, so full of promise, that all restraint was put aside, and every one abandoned himself to the pleasures of speculation and experiment. Would but every new and worthy idea find such an echo! One single year would then equal a century of progress. A delight it would be to live in such age, but a discoverer I would not wish to be.

Amongst the facts, which I have had the honor to bring to notice, is one claiming a large share of scientific interest, as well as of practical importance. I refer to the demonstration of the property of reflection, on which I have dwelt briefly. '

Having had opportunities to make many observations during my experience with vacuum bulbs and tubes, which could not be accounted for in any plausible way on any theory of vibration as far as I could judge, I began these investigations — disinclined, but expectant to find that the effects produced are due to a stream of material particles. I had many evidences of the existence of such streams. One of these I mentioned, describing the method of electrically exhausting a tube. Such exhaustion, I have found, takes place much quicker when the glass is very thin than when the walls are thick, I presume because of the easier passage of the ions. While a few minutes are sufficient when the glass is very thin, it often takes half an hour or more if the glass be thick or the electrode very large. In accordance with this idea I have, with a view of obtaining the most efficient action, selected the apparatus, and have found at each step my supposition confirmed and my conviction strengthened.

A stream of material particles, possessing a great velocity, must needs be reflected, and I was therefore quite prepared — assuming my original idea to be true — to demonstrate sooner or later this property. Considering that the reflection should be the more complete the smaller the angle of incidence, I adopted from the outset of my investigations a tube or bulb b of the form shown in

Fig. 1. It was made of very thick glass, with a bottom blown as thin as possible, with the two obvious objects of restricting the radiation to the sides and facilitating the passage through the bottom. A single electrode e, in the form of a round disk of a diameter slightly less than that of the tube, was placed about an inch below the narrow neck n on the top. The leading-in conductor c was provided with a long wrapping w, so as to prevent cracking, by the formation of sparks at the point where the wire enters the bulb. It was found advantageous for a number of reasons to extend the wrapping a good distance beyond the neck, on the inside and outside as well, and to place the seal-off in the narrow neck. On other occasions I have dwelt on the employment of an electrostatic screen in connection with such single-terminal bulbs. In the present instance the screen was preferably formed by a bronze paintings, slightly above the aluminum electrode and extending to just a little below the wrapping of the wire, so as to allow seeing constantly the end of the wrapping. Or else a small aluminum plate s, Fig. 2, was

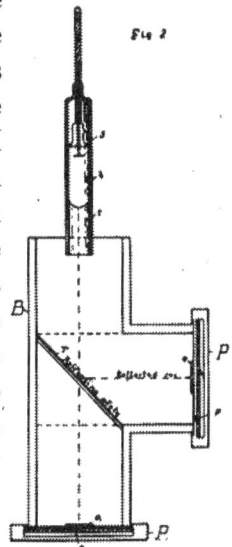

supported in the inside of the bulb above the electrode. This static screen practically doubles the effect, as it prevented all action above it. Considering, further, that the radiation sideways was restricted by the use of a very thick glass and most of it was thrown to the bottom by reflection, as I then surmised, it became evident that such a tube should prove much more efficient than one of ordinary form. Indeed, I quickly found that its power upon the sensitive plate was very nearly four times as great as that of a spherical bulb with an equivalent area of impact. This kind of tube is also very well adapted for use with two terminals by placing an external electrode e_1 as indicated by the dotted lines in Fig. 1. When the glass is taken thick the stream is sensibly parallel and concentrated. Furthermore, by making the tube as long as one desired, it was possible to employ very high potentials, otherwise impracticable with short bulbs.

The use of high potentials is of great importance, as it allows shortening considerably the time of exposure, and affecting the plate at much greater distances. I am endeavoring to determine more exactly the relation of the potential to the effect produced upon the sensitive plate. I deem it necessary to remark that the electrode should be of aluminum, as a platinum electrode, which is still persistently employed, gives inferior results and the bulb is disabled in comparatively short time. Some experimenters might find trouble in maintaining a fairly constant vacuum, owing to a peculiar process of absorption in the bulb, which has been pointed out early by Crookes, in consequence of which, by continued use, the vacuum may increase. A convenient way to prevent this I have found to be the following: The screen or aluminum plate s, Fig. 2, is placed directly upon the wrapping of the leading-in conductor c, but some distance back from the end. The right distance can be only determined by experience. If it is properly chosen, then, during the action of the bulb, the wrapping gets warmer, and a small bright spark jumps from time to time from the wire c to the aluminum plate s through the wrapping w. The passage of this spark causes gases to be formed; which slightly impair the vacuum; and in this manner, by a little skillful manipulation, the proper vacuum may be constantly maintained. Another way of getting the same result in a tube shown in Fig. 1 is to extend the wrapping so far inside that, when the bulb is' normally working, the wrapping is heated sufficiently to free gases to the required amount. It is for this purpose convenient to let the screen of bronze painting .*s* extend just a little below the wrapping, so that the spark may be observed. There are, however, many other ways of overcoming this difficulty, which may cause some annoyance to those working with inadequate apparatus:

In order to insure the best action the experimenter should note the various stages which I have pointed out before, and through which the bulb has to pass during the process of exhaustion. He will first observe that when the Crookes phenomena show themselves most prominently there is a reddish streamer issuing from the electrode, which in the beginning covers the latter almost entirely. Up to this point the bulb practically does not affect the sensitive plate, although the glass is very hot at the point of impact. Gradually the reddish streamer disappears, and just before it ceases to be visible the bulb begins to show better action; but still the effect upon the plate is very weak. Presently a white or even bluish stream is observed, and after some time the glass on the bottom of the bulb gets a glossy appearance. The heat is still more intense and the phosphorescence through the entire bulb is extremely brilliant: One should think that such a bulb must be effective, but appearances are often deceitful, and the beautiful bulb still does not work. Even when the white. or bluish stream ceases, and the glass on the bottom is so hot as to be nearly melting, the effect on the plate is very weak. But at this stage there appears suddenly at the bottom, of the tube a star-shaped changing design, as if the electrode would throw off drops of liquid. From this moment on the power of the bulb is tenfold, and at this stage it must always be kept to give the best results:

I may remark, however, that while it may be generally stated the Crookes vacuum is not high enough for the production of the Roentgen phenomena, this is not literally true. Nor are the Crookes phenomena produced at a particular degree of exhaustion, but manifest themselves even with poor vacua, provided the potential is high enough. This is likewise true of the Roentgen effects. Naturally, to verify this, provision must be made not to overheat the bulb when the potential is raised. This is easily done by reducing the number of impulses or their duration, when raising the potential: For such experiments, it will be found of advantage to use in connection with the ordinary induction coil a rotating commutator, instead of a vibrating brake. By changing the speed of the commutator, and also regulating the duration of contact, one is enabled to adjust the conditions to suit the degree of vacuum and potential employed.

In my experiments on reflection, presently considered, I have used the apparatus shown in Fig: 2. It consists of a T-shaped box throughout; of a square cross-section. The walls are mine of lead over one-eighth of an inch thick, which, under the conditions of the experiments, was found to be entirely impervious, even by long exposures to the rays. On the top end was supported firmly the bulb b, inclosed in a glass tube t of thick Bohemian glass; which reached some distance into the lead box. The lower end of the box was tightly closed by a plate-holder P_1, containing the sensitive film p_1, protected as usual.. Finally the side end was closed by a similar plate-holder P, with the sensitive protected film p. To obtain sharp images the objects o and o_1, exactly alike, were placed in the center of the fiber cover, protecting the sensitive plates. In the central portion of the box, provision was made for inserting a plate r of material; the reflective power of which was to be tested; and the dimensions of the box were such that the reflected ray and the direct one had to go through the same distance, the reflecting plate being at an angle of 45 degrees to the incident as well as reflected ray. Care was taken to exclude all possibility of action upon the plate p, except by reflected rays, and the reflecting plate r was made to fit tight all around in the lead box, so that no rays could reach the film p_1, except by passing through the plate to be tested. In my earliest experiments on reflection I observed only the effects of reflected rays, but in this instance, on the suggestion of Prof, Wm. A. Anthony, I provided the above means for simultaneously examining the action of the direct rays, which eventually passed through the reflecting plate. In this manner it was possible to compare the amount of .the transmitted and reflected radiation. The glass tube t surrounding the bulb b served to render the stream parallel and more intense. By taking impressions at various distances I found that through a considerable distance there was but little spreading of the bundle of rays or stream of particles.

To reduce the error which is caused unavoidably by too long exposures and very small distances, I reduced the exposure to an hour, and the total distance through which the rays had to pass before reaching the sensitive plates was 20 inches, the distance from the bottom of the bulb to the reflecting plate being 13 inches.

It is needless to remark that all the precautions in regard to the sensitive plates — constancy of potential, uniform working of the bulbs, and maintenance of the same conditions in general during these tests have been taken, as far as it was practicable. The plates to be tested were made of uniform size, so as to fit the space provided in the lead box. Of the conductors the following were tested: Brass, toolsteel, zinc, aluminum copper, lead, silver, tin;

and nickel, and of the insulators, lead-glass, ebonite, and mica. The summary of the observations is given in the following table:

Reflecting body	Impression by transmitted rays	Impression by reflected rays
Brass.	Strong.	Fairly strong.
Toolsteel.	Barely perceptible.	Very feeble.
Zinc.	None.	Very strong.
Aluminum.	Very strong.	None.
Copper.	None.	Fairly strong, but much less than zinc.
Lead.	None.	Very strong, but a little weaker than zinc.
Silver.	Strong, a thin plate being used.	weaker than copper.
Tin.	None.	Very strong; about like lead.
Nickel:	None.	About like copper.
Lead-glass.	Very strong.	Feeble.
Mica.	Very strong.	Very strong; about like lead.
Ebonite.	Strong.	About like copper.

By comparing, as in previous experiments, the intensity of the impression by reflected rays with an equivalent impression due to a direct exposure of the same bulb and at the same distance — that is, by calculating from the times of exposure under assumption that the action upon the plate was proportionate to the time — the following approximate results were obtained:

Reflecting body	Impression, by direct action	Impression by reflected rays
Brass	100	2
Toolsteel	100	0,5
Zinc	100	3
Aluminum	100	0
Copper	100	2
Lead	100	2,5
Silver	100	1,75
Tin	100	2,5
Nickel	100	2

Lead-glass	100	1
Mica	100	2,5
Ebonite	100	2

While these figures can be but rough approximations, there is, nevertheless, a fair probability that they are correct, in so far as the relative values of the impressions by reflected rays for the various bodies are concerned. Arranging the metals according to these values, and leaving for the moment the alloys or impure bodies out of question, we arrive at the following order: Zinc, lead, tin, copper, silver. The tin appears to reflect fully as well as lead, but, allowing for an error in the observation, we may assume that it reflects less, and in this case we find that this order is precisely the contact series of metals in air. If this proves true we shall be confronted with the most extraordinary fact. Why is zinc, for instance, the best reflector among the metals tested and why, at the same time, is it one of the foremost in the contact series? I have not as yet tried magnesium. The truth is that I was somewhat excited over these results. Magnesium should be even a better reflector than zinc, and sodium still better than magnesium. How can this singular relationship be explained? The only possible explanation seems to me at present that the bulb throws out streams of matter in some primary condition, and that the reflection of these streams is dependent upon some fundamental and electrical property of the metals. This would seem to lead to the inference that these streams must be of uniform electrification; that is, that they must be anodic or cathodic in character, but not both. Since the announcement, I believe in France for the first time, that the streams are anodic, I have investigated the subject and find that I can not agree with this contention. On the contrary, I find that anodic and cathodic streams both affect the plate, and, furthermore, I have been led to the conviction that the phosphorescence of the glass has nothing whatever to do with the photographic impressions. An obvious proof is that such impressions are produced with aluminum vessels when there is no phosphorescence, and, as regards the anodic or cathodic character, the simple fact that we can produce impressions by a luminous discharge excited by induction of a closed vessel, when there is neither anode nor cathode, would seem to dispose effectually of the assumption that the streams are issuing solely from one of the electrodes. It may, perhaps, be useful to point out here a simple fact in relation to the induction coils, which may lead an experimenter into an error. When a vacuum tube is attached to the terminals of an induction coil, both of the terminals are acted upon alike as long as the tube is not very highly exhausted. At a high degree of exhaustion both the electrodes act practically independently, and since they behave as bodies possessing considerable capacity, the consequence is that the coil is unbalanced. If the cathode, for instance, is very large, the pressure on the anode may rise considerably, and if the latter is made smaller, as is frequently the case, the electric density may be many times that on the cathode. It results from this that the anode gets very hot, while the cathode may be cool. Quite the opposite occurs if both of them are made exactly alike. But assuming the above conditions to exist, the hotter anode emits a more intense stream than the cool cathode, since the velocity of the particles is dependent on the electrical density, and likewise on the temperature.

From the previous tests air interesting observation can also be made in regard to the opacity. Far instance, a brass plate one-sixteenth inch thick proved fairly transparent while plates of zinc and copper of the same thickness showed themselves to be entirely opaque.

Since I have investigated reflection and arrived to results in this direction, I have been able to produce stronger effects by employing proper reflectors. By surrounding a bulb with a very thick glass tube the effect may be augmented very considerably. The employment of a zinc reflector in one instance showed an increase of about 40 per cent in the impression produced. I attach great practical value to the employment of proper reflectors, because by means of them we can employ any quantity of bulbs, and so produce any intensity of radiation required.

One disappointment in the course of these investigations has been the entire failure of my efforts to demonstrate refraction. I have employed lenses of all kinds and tried a great many experiments, but could not obtain any positive result.

On Roentgen Radiations
Electrical Review — April 8, 1896

Having observed the unexpected behavior of the various metals in regard to the reflection of these radiations, (see Electrical Review of April 1, 1896) I have endeavored to settle several still doubtful points. As, for the present, it appeared. chiefly desirable to establish the exact order of the metals, or conductors, in regard to their powers of reflection, leaving for further investigation the determination of the magnitude of the effects, I modified slightly the apparatus and procedure described in my communication just referred to. The reflecting plates were not made each of one metal, as before, but of two metals, the reflective power of which was to be compared. This was done by fastening upon a plate of lead the two metal plates to be investigated, so that the reflecting surface was divided in two halves by the joining line. Furthermore, to prevent any spreading and mingling of the rays reflected from both halves, I divided the lead box into two compartments by a thick lead plate through the middle. Care was taken that the density of the rays falling upon the reflecting surfaces was as uniform as possible, arid with this object in view the glass tube surrounding the bulb was lifted up so as to just expose the half-spherical bottom of the latter. The bulb was placed as exactly as it was practicable in the center, so that both halves of the reflecting plate were equally exposed to the radiations.

Having failed to obtain, in former experiments, a record for iron owing to an oversight, I tried to ascertain its position in the series by comparing it with copper, using a plate made up of iron and copper. The experiments showed that iron reflected about as much as copper, but which metal reflected better was impossible to determine with safety by this method. Next I endeavored to find whether tin or lead was a better reflector, by the same method. Three experiments were performed, and in each case the metals behaved nearly alike, but tin appeared just a trifle better. Finally I investigated the properties of magnesium as compared with zinc. In fact, the experiments showed that magnesium reflected a little better.

I am not yet satisfied, in view of the importance of this relation of the metals with the means employed, and will try to devise an apparatus which will do away with all the defects of the present. The time of the exposure I have found practicable to reduce to a few minutes by the help of a fluorescent paper.

In my previous communications I have barely hinted at the practical importance of the use of suitable reflectors. One would be apt to conclude that, since under the conditions of the previously described experiments, zinc, for instance, reflected only three per cent of the incident rays, the gain secured by the employment of such zinc reflector would be small. This, of course, would be an erroneous conclusion. First of all it should be remembered that in the instances mentioned before the angle of incidence was 45 degrees, and that for larger angles a much greater portion of the .rays would be reflected. The exact law of reflection is still to be determined. Now, let us suppose the shadow of an object is, taken at a distance, D. In order to get a sharp shadow we must take this distance not less than two feet, and I am finding it more and more necessary to adopt a still greater distance. If, for the sake of simplicity, we assume a spherical bulb and electrode, the radiation will be uniform on all sides, and any element of a surface of a sphere of radius D, drawn around the electrode, will receive an equal quantity of rays. The total surface of this sphere will be 4 pi D^2. The object, the shadow of which is to be taken, may have a small area a, which gets only an insignificant part of the total rays emitted, this part being given by the proportion $\frac{a}{4\pi D^2}$. In reality we can not assume less than $\frac{a}{D^2\pi}$ a/D^2pi as effective ratio, but even then, if D is very large and the object, that is, the area a, small, this ratio may be still so small that evidently, by the use of a proper reflector, we can, easily concentrate upon the area a an amount of rays several times exceeding that which would fall upon it without the use of a reflector, in spite of the fact that we are able to reflect only a few per cent of the total incident rays.

As an evidence of the effectiveness of such a reflector, the annexed print of the shoulder and ribs of a man is shown. A funnel-shaped zinc reflector, two feet high, with an opening of five inches on the bottom and 23 inches at the top, was used in the experiment. A tube, similar in every respect to those previously described, was suspended in the funnel, so that just the static screen of the tube was above the former. The exact distance from the electrode to the sensitive plate was four and one-half feet. The distance from the end of the tube to the plate was three and one-half feet. The exposure lasted 40 minutes. The plate showed very strongly and clearly every bone, and shoulder and ribs, but I can not tell how clearly they will appear in the print. I selected the same object as in my first report in your columns on this investigation, so as to give a better idea of the progress made. The advance will be best appreciated by stating that the distance in this case was much more than double, while the time of exposure was less than one-half. The chief importance of a reflector consists, however, in this, that it allows the use of many bulbs without sacrifice of precision and clearness, and also the concentration of a great quantity of radiation upon a very small area.

Since the use of phosphorescent or fluorescent bodies in connection with the sensitive film has been suggested by Professors Henry and Salvioni, I have found it an easy matter to shorten the time of exposure to a few minutes, or even seconds. So far, it seems that the tungstate of calcium, recently introduced by Edison, and manufactured by Messrs. Aylsworth & Jackson, is the most sensitive body. I obtained a sample of it and used it in a series of tests. It fluoresces decidedly better than barium-platino-cyanide, but, owing to the size of the crystals and necessarily uneven distribution on the paper, it does not leave a dean impression. For use in connection with the sensitive films, it should be ground very fine, and some way should be adopted of distributing it uniformly. The paper also must adhere firmly to the film all over the plate, so as to get fairly sharp outlines. The fluorescence of this body seems to depend on a peculiar radiation, because I tested several bulbs, which otherwise worked excellently, without producing a very good result, and I almost gained a false impression.. One or two of the bulbs, however, effected it very powerfully. An impression of the hand was taken at a distance of about six feet from the bulb with an exposure of less khan one minute, and even then it was found that the plate was overexposed. I then took an impression of the chest of a man at a distance of 12 feet from the end of the tube, exposing five minutes. The developed plate showed the ribs dearly, but the outlines were not sharp. Next, I employed a tube with a zinc reflector, as before described, taking an impression of the chest of an assistant at a distance of four feet from the bulb. The latter was strained a little too much in this experiment and exploded, in consequence of the great internal pressure against the bombarded spot. This . accident will frequently occur when the bulb is strained too, high, the preceding outward sign being an increased activity and vapor like appearance of the gas in the bulb and rapid heating of the latter. The process causing the abnormal increase in the internal pressure against the glass wall seems to be due to some action opposite to that noted by Crookes and Spottiswoode, and is very rapid, and for this reason the experimenter should watch carefully for these ominous signs and instantly reduce the potential. Owing to the

untimely end of the bulb in this last described experiment, the exposure
lasted only one minute. Nevertheless, a very strong impression of the
skeleton of the chest, showing the right and left ribs and other details,
was obtained. The outlines, however, were again much less sharp than
when the ordinary process without the phosphorescent intensifier was
followed, although care was taken to press the fluorescent paper firmly
against the film. From the foregoing it is evident that, when using the
above means for shortening the time of exposure, the thickness of the
object is not of very much consequence.

I obtained a still better idea of the quality of tungstate of calcium by
observing the effect upon a fluorescent screen made of this chemical.
Such a screen, together with a paper box, has been termed with the
fanciful name "fluoroscope." It is really Salvioni's Cryptoscope with the
lens omitted, which is a great disadvantage. To appreciate the
performances of such a screen, it is necessary to work at night, when
the eye has for a long time been used to the darkness, and made
capable of noting the faint effects on the screen. In one instance the
performance of this screen was particularly noteworthy. It was
illuminated at a distance of 20 feet, and even at a distance of 40 feet I
could still observe a faint shadow passing across the field of vision,
when moving the hand in front of the instrument. Looking at a distance
of about three feet from the bulb through the body of an assistant, I
could distinguish easily the spinal column in the upper part of the body,
which was more transparent. In the lower part of the body the column
and the rest were practically not perceptible. The ribs were only very
faintly seen. The bones of the neck were plainly noticeable, and I could
see through the body of the assistant very easily a square plate of
copper, as it was moved up and down in front of the bulb. When looking
through the head I could observe only the outline of the skull and the
chin-bone, yet the field of vision was still bright. Everything else
appeared indistinct. This shows that improving of fluorescence will not
aid us very much in the examination of the internal parts of the body.
The solution rather will come through the production of very powerful
radiations, capable of producing very strong. shadows. I believe I have
indicated the right way to secure this result. Although it must be
admitted that the performance of such a screen is remarkable with the
appliances I have used, I have, nevertheless, convinced myself of its still
limited value for the purpose of examination. We can distinguish the
bones in the limbs, but not nearly as clearly as a photographic
impression shows it. Eventually, however, with the help of strong
radiations and good reflectors, such fluorescent screens may become
valuable instruments for investigation. A few weeks ago, when I
observed a small screen of barium-platino-cyanide flare up at a great
distance from the bulb, I told some friends that it might be possible to

observe by the aid of such a screen objects passing through a street. This possibility seems to me much nearer at present than it appeared then. Forty feet is a fair width for a street, and a screen lights up faintly at that distance from a single bulb. I mention this odd idea only as an illustration of how these scientific developments may even affect our morals and customs. Perhaps we shall shortly get so used to this state of things that nobody will feel the slightest embarrassment while he is conscious that his skeleton and other particulars are being scrutinized by indelicate observers.

Fluorescent screens afford some help in getting an idea of the condition of the bulb when working. I hoped to find some evidence of refraction by means of such a screen, placing a lens between it and the bulb, and varying the focal distance. To my disappointment, although the shadow of the lens was observable at a distance of 20 feet, I could see no trace of refraction. The use of. the screen for the purpose of noting the effects of reflection and diffraction proved likewise futile.

Roentgen Ray Investigations
Electrical Review — April 22, 1896

Further investigations concerning the behavior of the various metals in regard to reflection of these radiations have given additional support to the opinion which I have before expressed; namely, that Volta's electric contact series in air is identical with that which is obtained when arranging the metals according to their powers of reflection, the most electro-positive metal being the best reflector. Confining myself to the metals easily experimented upon, this series is magnesium, lead, tin, iron, copper, silver, gold and platinum. The last named metal should be found to be the poorest, and sodium one of the best, reflectors. This relation is rendered still more interesting and suggestive when we consider that this series is approximately the same which is obtained when arranging the metals according to their energies of combination with oxygen, as calculated from their chemical equivalents.

Should the above relation be confirmed by other physicists, we shall be justified to draw the following conclusions: *First,* the highly exhausted bulb emits material streams which, impinging on a metallic surface, are reflected; *second,* these streams are formed of matter in some primary or elementary condition; *third,* these material streams are probably the same agent which is the cause of the electro-motive tension between metals in close proximity or actual contact, and they may possibly, to some extent, determine the energy of combination of the metals with oxygen; *fourth,* every metal or conductor is more or less a source of such streams; *fifth,* these streams or radiations must be produced by some radiations which exist in the medium; and *sixth,* streams resembling the cathodic must be emitted by the sun and probably also by other sources of radiant energy, such as an arc light or Bunsen burner.

The first of these conclusions, assuming the above-cited fact to be correct, is evident and incontrovertible. No theory of vibration of any kind would account for this singular relation between the powers of reflection and electric properties of the metals, Streams of projected matter coming in actual contact with the reflecting' metal surface afford the only plausible explanation.

The second conclusion is likewise obvious, since no difference whatever is observed by employing various qualities of glass for the bulb, electrodes of different metals and any kind of residual gases. Evidently, whatever the matter constituting the streams may be, it must undergo a change in the process of expulsion, or, generally speaking; projection — since the views in this regard still differ — in such a way as to lose entirely the characteristics which it possessed when forming the electrode, or wall of the bulb, or the gaseous contents of the latter.

The existence of the above relation between the reflecting and contact series forces us likewise to the third conclusion, because a mere coincidence of that kind is, to say the least, extremely improbable. Besides, the fact may be cited that there is always a difference of potential set up between two metal plates at some distance and in the path of the rays issuing from an exhausted bulb.

Now, since there exists an electric pressure of difference of potential between two metals in dose proximity or contact, we must, when considering all the foregoing, come to the fourth conclusion, namely, that the metals emit similar streams, and I therefore anticipate that, if a sensitive film be placed between two plates, say, of magnesium and copper, a true Roentgen shadow picture would be obtained after a very long exposure in the dark. Or, in general, such picture could be secured whenever the plate is placed near a metallic or conducting body, leaving for the present the insulators out of consideration. Sodium, one of the first of the electric contact series, but not yet experimented upon, should give out more of such streams than even magnesium.

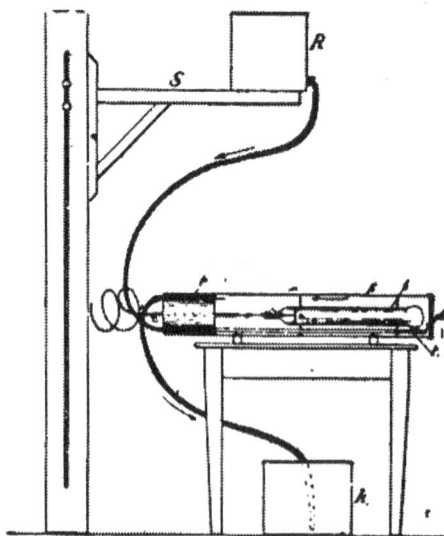

Obviously, such streams could not be forever emitted, unless there is a continuous supply of radiation from the medium in some other form; or possibly the streams which the bodies themselves emit are merely reflected streams coming from other sources. But since all investigation has strengthened the opinion advanced by Roentgen that for the production of these radiations some impact is aired, the former of the two possibilities is the more probable one, and we must assume that the radiations existing in the medium and giving rise to those here considered partake something of the nature of cathodic streams.

But if such streams exist all around us in the ambient medium, the question arises, whence do they come? The only answer is: From the sun. I infer, therefore, that the sun and other sources of radiant energy must, in a less degree, emit radiations or streams of matter similar to those thrown off by an electrode in a highly exhausted inclosure. This seems to be, at this moment, still a point of controversy. According to my present convictions a Roentgen shadow picture should, with very long exposures, be obtained from all sources of radiant energy, provided the radiations are permitted first to impinge upon a metal or other body.

The preceding considerations tend to show that the lumps of matter composing a cathodic stream in the bulb are broken up into incomparably smaller particles by impact against the wall of the latter, and, owing to this, are enabled to pass into the air. All evidence which I have so far obtained points rather to this than to the throwing off of particles of the wall itself under the violent impact of the cathodic stream. According to my convictions, then, the difference between Lenard and Roentgen rays, if there be any, lies solely in this, that the particles composing the latter are incomparably smaller and possess a higher velocity. To these two qualifications I chiefly attribute the non-deflectibility by a magnet which I believe will be disproved in the end. Both kinds of rays, however, affect the sensitive plate and fluorescent screen, only the rays discovered by Roentgen are much more effective. We know now that these rays are produced under certain exceptional conditions in a bulb, the vacuum being extremely high, and that the *range* of greatest activity is rather small.

I have endeavored to find whether the reflected rays possess certain distinctive features, and I have taken pictures of various objects with this purpose in view, but no marked difference was noted in any case. I therefore conclude that the matter composing the Roentgen rays does not suffer further degradation by impact against bodies. One of the most important tasks for the experimenter remains still to determine what becomes of the energy of these rays. In a number of experiments with rays reflected from and transmitted through a conducting of insulating plate, I found that only a small part of the rays could be accounted for. For instance, through a zinc plate, one-sixteenth of an inch thick, under an incident angle of 45 degrees, about two and one-half per cent were reflected and about three per cent transmitted through the plate, hence over 94 per cent of the total radiation remain to be accounted for. All the tests which I have been able to make have confirmed Roentgen's statement that these rays are incapable of raising the temperature of a body. To trace this lost energy and account for it in a plausible way will be equivalent to making a new discovery..,

Since it is now demonstrated that all bodies reflect more or less, the diffusion through the air is easily accounted for. Observing the tendency to scatter through the air, I have been led to increase the efficiency of reflectors by providing not one; but separated successive layers for reflection, by making the reflector of thin sheets of metal; mica or other substances. The efficiency of mica. as a reflector I attribute chiefly to the fact that it is composed of many superimposed layers which reflect individually. These many successive reflections are, in my opinion, also the cause of the scattering through the air.

In my communication to you of April 1, I have for the first time stated that these rays are composed of matter in a "primary" or elementary condition or state. I have chosen this mode of expression in order to avoid the use of the word "ether," which is usually understood in the sense. of the Maxwellian interpretation, which would not be in accord with my present convictions in regard to the nature of the radiations.

An observation which might be of some interest is the following: A few years ago I described on one occasion a phenomenon observed in highly exhausted bulbs. It is a brush or stream issuing from a single electrode under certain conditions, which rotates very rapidly in consequence of the action of the earth's magnetism. Now I have recently observed this same phenomenon in several bulbs which were capable of impressing the sensitive film and fluorescent screen very. strongly. As the brush is rapidly twirling around I have conjectured that perhaps also the Lenard and Roentgen streams axe rotating under the action of the earth's magnetizing and I am endeavoring to obtain an evidence of such motion by studying the action of a bulb in various positions with respect to the magnetic axis of the earth.

In so far as the vibrational character of the rays is concerned, I still hold that the vibration is merely that which is conditioned by the apparatus employed. With the ordinary induction coil we have almost exclusively to deal with a very low vibration impressed by the commutating device or brake. With the disruptive coil we usually have a very strong superimposed vibration in addition to the fundamental one, and it is easy to trace sometimes as much as the fourth octave of the fundamental vibration. But I can not reconcile myself with the idea of vibrations approximating or even exceeding those of light, and think that all these effects could be as well produced with a steady electrical pressure as from a battery, with the exclusion of all vibration which may, occur, even in such instance, as has been pointed out by De La Rive. In my experiments I have tried to ascertain whether a greater difference between the shadows of the bones and flesh could be obtained by employing currents of extremely high frequency, but I have been unable to discover any such effect which would be dependent on the frequency of the currents, although the latter were varied between as wide limits as :was possible. But it is a rule that the more intense the action the .sharper the shadows obtained, provided that the distance is not too small. It is furthermore of the greatest importance for the clearness of the shadows that the rays should be passed through some tubular reflector, which renders them sensibly parallel.

In order then to bring out as much detail as possible on a sensitive plate, we have to proceed in precisely the same way as if we had to deal with flying bullets hitting against a wall composed of parts of different density with the problem before us of producing as large as possible a difference in the trajectories of the bullets which pass through the various parts of the wall. Manifestly, this difference will be the greater the greater the velocity of the bullets; hence, in order 'to bring out detail, very strong radiations are required. Proceeding on this theory I have employed exceptionally thick films and developed very slowly, and in this way clearer pictures have been obtained. The importance of slow development has been first pointed out by Professor Wright, of Yale. Of course, .if Professor Henry's suggestion of the use of a fluorescent body in contact with the sensitive film is made use of, the process is reduced to an ordinary quick photographic procedure, and the above consideration does not apply.

It being desirable to produce as powerful a radiation as possible, I have continued to devote my attention' to this problem and have been quite successful. First of all, there existed limitations in the vacuum tube which did not permit the applying of as high a potential as I desired; namely, when a certain high degree of exhaustion was reached a spark would form behind the electrode, which would prevent straining the tube much higher. This inconvenience I have overcome entirely by making the wire leading to the electrode very long and passing it through a narrow channel, so that the heat from the electrode could not cause the formation of such sparks. Another limitation was imposed by streamers which would break out at the end of the tube when the potential was excessive. This latter inconvenience I have overcome

either by the use of a cold blast of air along the tube, as I have
mentioned before, or else by immersion of the tube in oil. The oil, as it
is now well known, is a means of rendering impossible the formation of
streamers by the exclusion of all air. The use of the oil in connection
with the production of these radiations has been early advocated in this
country by Professor Trowbridge. Originally I employed a wooden box
made thoroughly tight with wax and filled with oil or other liquid, in
which the tube was immersed. Observing certain specific actions, I
modified and improved the apparatus, and in my later investigations I
have employed an arrangement as shown in the annexed cut. A bulb b,
of the kind described before, with a leading-in wire and neck much
longer than here shown, was, inserted into a large and thick glass tube
t. The tube was closed in front by a diaphragm d of pergament, and by
a rubber plug P in the back. The plug was provided with two holes, into
the lower one of which a glass tube t_1, reaching to very nearly the end
of the bulb, was inserted. Oil of some kind was made to flow through
rubber tubes r r from a large reservoir R, placed on an adjustable
support S, to the lower reservoir R_1, the path of the oil being clearly
observable from the drawing. By adjusting the difference of the level
between the two reservoirs it was easy to maintain a permanent
condition of working. The outer glass tube t served in part as a reflector,
while at the same time it permitted the observation of the bulb b during
the action. The plug P, in which the conductor c was tightly sealed, was
so arranged that it could be shifted in and out of the tube t, so as to vary
the thickness of the oil traversed by the rays.

I have obtained some results with this apparatus which clearly show
the advantage of such disposition. For instance, at a distance of 45 feet
from the end of the bulb my assistants and myself could observe clearly
the fingers of the hand through a screen of tungstate of calcium, the
rays traversing about two and one half inches of oil and the diaphragm
d. It is practicable with such apparatus to make photographs of small
objects at a distance of 40 feet, with only a few minutes exposure, by the
help of Professor Henry's method. But, even without the use of a
fluorescent powder, short exposures are practicable, so that I think the
use of the above method is not essential for quick procedure. I rather
believe that in the practical development of this principle, if it shall be
necessary, Professor Salvioni's suggestion of a fluorescent emulsion,
combined with a film, will have to be adopted. This is bound to give
better results than an independent fluorescent screen, and will very
much simplify the process. I may say, however, that, since my last
communication, considerable improvement has been made in the

screens. The manufacturers of Edison's tungstate of calcium are now furnishing screens which give fairly clean pictures. The powder is fine and it is more uniformly distributed. I consider, also, that the employment of a softer and thicker paper than before is of advantage. It is just to remark that the tungstate of calcium has also proved to be an excellent fluorescent in the bulb. I tested its qualities for such use immediately and find it so "far unexcelled. Whether it will be so for a long time remains to be seen. News reaches us that several fluorescent bodies, better than the cyanides, have been discovered abroad.

Another improvement with a view of increasing the sharpness of the shadows has been proposed to me by Mr. E. R. Hewitt. He assumed that the absence of sharpness of the outlines in the shadows on the screen was due to the spread of the fluorescence frown crystal to crystal. He proposed to avoid this by using a thin aluminum plate with many parallel .grooves. Acting on this suggestion, I made some experiments with wire gauze and, furthermore, with screens made of a mixture of a fluorescent with a non-fluorescent powder. T found that the general brightness of the screen was diminished, but that with a strong radiation the shadows appeared sharper. This idea might be found capable of useful application.

By the use of the above apparatus I have been enabled to examine much better than before the body by means of the fluorescent screen. Presently the vertebral column can be seen quite clearly, even in the lower part of the body. I have also clearly noted the outlines of the hip bones. Looking in the region of the heart I have been able to locate in unmistakably. The background appeared much brighter, and this difference in the intensity of the shadow and surrounding has surprised me. The ribs I could now see on a number of occasions quite distinctly, as well as the shoulder bones. Of course, there is no difficulty whatever in observing the bones of all limbs. I noted certain peculiar effects which I attribute to the oil. For instance, the rays passed through plates of metal over one-eighth of an inch thick, and in one instance I could see quite clearly the bones of my hand through sheets of copper, iron and brass of a thickness of nearly' one-quarter of an inch. Through glass the rays seemed to pass with such freedom that, looking through the screen in a direction at right angles to the axis of the tube, the action was most intense, although the rays had to pass through a great thickness of glass and oil. A glass slab nearly one-half of an inch thick, held in front of the screen, hardly dimmed the fluorescence. When holding the. screen in front of the tube at a distance of about three feet, the head of an assistant, thrust between the screen and the tube, cast but a feeble shadow. It appeared some times as if the bones and the flesh were equally transparent to the radiations passing through, the oil. When very close to the bulb, the screen was illuminated through the body of an assistant so strongly that, when a hand was moved-in, front, I could clearly note the motion of the hand. through the body. In one instance I could even distinguish the bones of the arm.

Having observed the extraordinary transparence of the bones in some instances, I at first surmised that the rays might be vibrations of high pitch, and that the oil had in. some way absorbed a part of them. This view, however, became untenable when I found that at a certain distance from the bulb I obtained a sharp shadow of the bones. This latter observation led me to apply usefully the screen in taking impressions on the plate. Namely, in such ,case it is of advantage to first determine by means of the screen the proper distance at which the object is to be placed before taking the impression. It will be found that often the image is much clearer at a greater distance. In order, to avoid any error when observing with the screen, I have surrounded the box with thick metal plates, so as to prevent the fluorescence, in consequence of the radiations, reaching the screen from the sides. I believe that such an arrangement is absolutely necessary if one wishes to make correct observations.

During my study of the behavior of oils and other liquid insulators, which I am still continuing, it has occurred to me to investigate the important effect discovered by Prof. J. J. Thomson. He announced some time ago that all bodies traversed by Roentgen radiations become conductors of electricity. I applied a sensitive resonance test to the investigation of this phenomenon in a manner pointed out in my earlier writings on high frequency currents. A secondary, preferably not in very close inductive relation to the primary circuit, was connected to the latter and to the ground, and the vibration through the primary, was so adjusted that true resonance took place. As the secondary had a considerable number of turns, very small bodies attached to the free terminal produced considerable variations of potential on the latter. Placing a tube in a box of wood filled with oil and attaching it to the terminal, I adjusted the vibration through the primary so that resonance took place without the bulb radiating Roentgen rays to an appreciable extent. I then changed the conditions so that the bulb became very active in the production of the rays. The oil should have now, according to Prof. J. J. Thomson's statement, become a conductor and a very marked change in the vibration should have occurred. This was found not to be the case, so that we must see in the phenomenon discovered by J. J. Thomson only a further evidence that we have to deal here with streams of matter which, traversing the bodies, carry away electrical charges. But the bodies do not become conductors in the common acceptance of the term. The method I have followed is so delicate that a mistake is almost an impossibility.

An Interesting Feature of X-Ray Radiations
Electrical Review — July 8, 1896

The following observations, made with bulbs emitting Roentgen radiations, may be of value in throwing additional light upon the nature of these radiations, as well as illustrating better properties already known. In the main these observations agree with the views which have forced themselves upon my, mind from the outset, namely, :that the rays consist of streams of minute material particles projected with great velocity. In numerous experiments I have found that the matter which, by impact within the bulb, causes the formation of the rays may come from tidier of the electrodes. Inasmuch as the latter are by continued use disintegrated to a marked degree, it seems more plausible to assume that the projected matter consists of parts of the electrodes themselves rather than of the residual gas. However, .other observations, upon which I can not dwell at present, lead to this conclusion. The lumps of projected matter are by impact further disintegrated into particles so minute as to be able to pass through the walls of the bulb: or else they tear off such particles from the walls, or generally bodies, against which they are projected. At any rate, an impact and consequent shattering seems absolutely necessary for the production of Roentgen rays. The vibration, if there be any, is only that which is impressed by the apparatus, and the vibrations can only be longitudinal.

The principal source of the rays is invariably the place of first impact within the bulb, whether it be the anode, as in some forms of tube, or an inclosed insulated body, or the glass wall. When the matter thrown off from an electrode, after striking against an obstacle, is thrown against another body, as the wall of the bulb, for instance, the place of second impact is a very feeble source of the rays.

These and other facts will be better appreciated by referring to the annexed, figure, in which a form of tube is shown used in a number of my experiments. The general form is that described on previous occasions. A single electrode *e*, consisting of a massive aluminum plate, is mounted on a conductor *t,* provided with a glass wrapping *w* as usual, and sealed in one of the ends of a straight tube *b*, about five centimeters in diameter and 30 centimeters long. The other end of the tube is blown out into a thin bulb of a slightly larger diameter, and near this end is supported on a glass stem a a funnel f of thin platinum sheet, In such bulbs I have used a number of different metals for impact with a view of increasing the intensity of the rays and also for the purpose of reflecting and concentrating them. Since, however, in a later contribution, Professor Roentgen has pointed out that platinum gives the most intense rays, I have used chiefly this metal, finding a marked increase in the effect upon the screen or sensitive plate. The particular object of the presently described construction was to ascertain whether the rays generated at the inner surface of the platinum funnel f would be brought to a focus outside of the bulb, and further, whether they would proceed in straight lines from that point. For this purpose the apex of the platinum cone was arranged to be about two centimeters outside of the bulb at o.

When the bulb was properly exhausted and set in action, the glass wall below the funnel f became strongly phosphorescent, but not uniformly, as there was a narrow ring r r on the periphery brighter than the rest, this ring being evidently due to the rays reflected from the platinum sheet.

Placing a fluorescent screen in contact or quite close to the glass wall below the funnel, the portion of the screen in the immediate neighborhood of the phosphorescent patch was brightly illuminated, the outlines being entirely indistinct. Receding now with the screen from the bulb, the strongly illuminated spot became smaller and the outlines sharper, until, when the point o was reached, the luminous part had dwindled down to a small point. Moving the screen a few millimeters beyond o caused a small dark spot to appear, which widened into a circle and became larger and larger in the same measure as the distance from the bulb was increased (see S), until, at a sufficiently large distance, the dark circle covered the entire screen. This experiment illustrated in a beautiful way the propagation in straight lines, which Roentgen originally proved by pinhole photographs. But, besides this, an important point was noted; namely, that the fluorescent glass wall emitted practically no rays, whereas, had the platinum not been present, it would have been, under similar conditions, an efficient source of the rays, for the glass, even by weak excitation of the bulb, was strongly heated. I can only explain the absence of the radiation from the glass by assuming that the matter proceeding from the surface of the platinum sheet was already in a finely divided state when it reached the glass wall. A remarkable fact is, also, that, at least by a weak excitation of the bulb, the edges of the dark circle were very sharp, which speaks strongly against diffusion. By exciting the bulb very strongly, the background became brighter and the shadow S fainter, though it continued to be plainly visible even then.

From the preceding it is evident .that, by a suitable construction of the bulb, the rays emanating from the latter may be concentrated upon any small area at some distance, and a practical advantage may be taken of this fact when producing impressions upon a plate or examining bodies by means of a fluorescent screen.

Roentgen Rays Or Streams

Electrical Review — August 12, 1896

In the original report of his epochal discoveries, Roentgen expressed his conviction that the .phenomena he observed were due to certain novel disturbances in the ether. This opinion deserves to be considered the more as it was probably formed in the first enthusiasm over the revelations, when the mind of the discoverer was capable of a much deeper insight into the nature of things.

It was known since long ago that certain dark radiations, capable of penetrating opaque bodies, existed, and when the rectilinear propagation, the action on a fluorescent screen and on a sensitive film was noted, an obvious and unavoidable inference was that the new radiations were transverse vibrations, similar to those known as light. On the other hand, it was difficult to resist certain arguments in favor of the less popular theory of material particles, especially as, since the researches of Lenard, it has become very probable that material streams, resembling the cathodic, existed in free air, Furthermore, I myself have brought to notice the fact that similar material streams — which were subsequently, upon Roentgen's announcement, found capable of producing impressions on a sensitive film — were obtainable in free air, even without the employment of a vacuum bulb, simply by the use of very high potentials, suitable for imparting to the molecules of the air or other particles a sufficiently high; velocity. In reality, such puffs or jets of particles are formed in the vicinity of every highly charged conductor, the potential of which is rapidly varying, and I have shown that, unless they are prevented, they are fatal to every condenser or high-potential transformer, no matter how thick the insulation. They also render practically valueless any estimate of the period of vibration of an electro-magnetic system by the usual made of calculation or measurement in a static condition in all cases in which the potential is very high and the frequency excessive.

It is significant that, with these and other facts before him, Roentgen inclined to the conviction that the rays he discovered were longitudinal waves of ether.

After a long and careful investigation, with apparatus excellently suited for the purpose, capable of producing impressions. at great distances, and after examining the results pointed out by other experimenters, I have come to the conclusion which I have already intimated in my former contributions to your esteemed journal, and which I now find courage to pronounce without hesitation, that the original hypothesis of Roentgen will be confirmed in two particulars; first, in regard to the longitudinal character of the disturbances; second, in regard to the medium concerned in their propagation. The present expression of my views is made solely for the purpose of preserving a faithful record of what, to my mind, appears to be the true interpretation of these new and important manifestations of energy.

Recent observations of some dark radiations from novel sources by Becquerel and others, and certain deductions of Helmholtz, seemingly. applicable to the explanation of the peculiarities of the Roentgen rays, have given additional weight to the arguments on behalf of the theory of transverse vibrations, and accordingly this interpretation of the phenomena is held in favor. But this view is still of a purely speculative character, being, as it is at present, unsupported by any conclusive experiment. Contrarily, there is considerable experimental evidence that some matter is projected with great velocity from the bulbs, this matter being in all probability the only cause of the actions discovered by Roentgen.

There is but little doubt at present that a cathodic stream within a bulb is composed of small particles of matter, thrown off with great velocity from the electrode. The velocity probably attained is estimable, and fully accountable for the mechanical and heating effects produced by the impact against the wall or obstacle within the bulb. It is, furthermore, an accepted view that the projected lumps of matter act as inelastic bodies, similarly to ever so many small lead bullets. It can be easily shown that the velocity of the stream *may* be as much as 100 kilometers a second, or even more, at least in bulbs with a single electrode, in which the practicable vacua and potentials are much higher than in the ordinary bulbs with two electrodes. But, now, matter moving with such great velocity must surely penetrate great thicknesses of the obstruction in its path, if the laws of mechanical impact are at all applicable to a cathodic stream. I have presently so much familiarized myself with this view that, if I had no experimental evidence, I would not question the fact that some matter is projected through the thin wall of a vacuum tube. The exit from the latter is, however,, the more likely to occur, as the lumps of matter must be shattered into still much smaller particles by the impact, From my experiments on reflection of the Roentgen rays, before reported, which, with powerful radiations, may be shown to exist under all angles of incidence, it appears that the lumps or molecules are indeed shattered into fragments or constituents so small as to make them lose entirely some physical properties possessed before the impact. Thus, the material composing the electrode, the wall of the bulb or obstruction of any kind placed within the latter, are of absolutely no consequence, except in so far as the intensity of the radiations is concerned. It also appears, as I have pointed out, that no further disintegration of the lumps is attendant upon a second impact. The matter composing the cathodic stream is, to all evidence, reduced to matter of some primary form, heretofore not known, as such velocities and such violent impacts have probably never been studied or even attained before these extraordinary manifestations were observed. Is it not possible that the very ether vortexes which, according to Lord Kelvin's ideal theory, compose the lumps, are dissolved, and that in the Roentgen phenomena we may witness a transformation of ordinary matter into ether? It is in this sense that,. I think, Roentgen's first hypothesis will be confirmed. In such case there can be, of course, no question of waves other than the longitudinal assumed by him, only, in my opinion, the frequency must be very small — that of the electro-magnetic vibrating system — generally not more than a few millions a second. If such process of transformation does take place, it will be difficult, if not impossible, to determine the amount of energy represented in the radiations, and the statement that this amount is very small should be received with some caution.

I As to the rays exhaustively studied by Lenard, which have proved to be the nucleus of these great realizations, I hold them to be true cathodic streams, projected through the wall of the tube. Their deflectibility by a magnet shows to my mind simply that they differ but little from those within the bulb. The lumps of matter are probably large and the velocity small as compared with .that of the Roentgen rays. They should, however be capable in a minor degree of all the actions of the latter. These actions I consider to, be purely :mechanical and obtainable by other means. So, for instance, I think that if a gun loaded with mercury were fired through a thin board, the projected mercury vapor would cast a shadow of an object upon a film made especially sensitive to mechanical impact, or upon a screen of material capable of being rendered fluorescent by such impact.

The following observations made by myself and others speak more or less for the existence of the streams of matter.

I — Phenomena of Exhaustion

On this subject I have expressed myself on another occasion. It is only necessary to once more point out that the effect observed by me should not be confounded with that noted by Spottiswoode and Crookes. I explain the latter phenomenon as follows: The first fluorescence appearing when the current is turned on, is due to some organic matter almost always introduced in the bulb in the process of manufacture. A minute layer of such matter on the wall produces invariably this first fluorescence, and the latter never takes place when the bulb has been exhausted under application of a high degree of heat or when the organic matter is otherwise destroyed. Upon the disappearance of the first fluorescence .the rarefaction increases slowly, this being . a necessary result of particles being projected from the electrode and fastening themselves upon the wall. These particles absorb a large portion of the residual gas. The latter can. be again freed by the application of heat to the bulb or otherwise. So much of the effects observed by these investigators. In the instance observed by myself, there must be actual expulsion of matter, and for this speak following facts: (a) the exhaustion is quicker when the glass is thin; (b) when the potential is higher; © when the discharges are more sudden; (d) when there is no obstruction within the bulb; (e) the exhaustion takes place quickest with an aluminum or platinum electrode, the former metal giving particles moving with greatest velocity, the latter particles of greatest weight; (f) the glass wall, when softened by the heat, does not collapse, but bulges outwardly; (g) the exhaustion takes place, in some cases, even if a small perceptible hole is pierced through the glass; (h) all causes tending to impart a greater velocity to the particles hasten the process of exhaustion.

II — Relation Between Opacity and Density

The important fact pointed out early by Roentgen and confirmed by subsequent research, namely, that a body is the more opaque to the rays the denser it is, can not be explained as satisfactorily under any other assumption as that of the rays being streams of matter, in which case such simple relation between opacity and density would necessarily exist. This relation is the more important in its bearing upon the nature of the rays, as it does not at all exist in light-giving vibrations, and should consequently not be found to so marked a degree and under *all conditions with vibrations,* presumably similar to and approximating in frequency the light vibrations.

III — Definition of Shadows on Screen or Plate
When taking impressions or observing shadows while varying the intensity of the radiations, but maintaining all other conditions as nearly as possible alike, it is found that the employment of more intense radiations secures little, if any, advantage, as regards the definition of the details. At first it was thought that all' there was needed was to produce very powerful rays. But the experience was disappointing, for, while I succeeded in producing rays capable of impressing a plate at distances of certainly not less than 30 meters, I obtained but slightly better results. There was one advantage in using such intense rays, and this was that the plate could be further removed from the source, and consequently a better shadow was obtained. But otherwise nothing to speak of was gained. The screen in the dark box would be at times rendered so bright as to allow reading at some distance plainly, but the shadow was not more distinct for all that. In fact, often a very strong radiation gave a poorer impression than a weak one. Now, a fact which I have repeatedly observed and to which I attach great importance in this connection, is the following: When taking impressions at a small distance with a tube giving very intense rays, no shadow, unless a scarcely perceptible one, is obtained. Thus, for instance, the flesh and bones of the hand appear equally transparent. Increasing presently gradually the distance, it is found that the bones cast a shadow, while the flesh leaves no impression. The distance still increased, the shadow of the flesh appears, while that of the bones grows deeper, and in this neighborhood a place can be found at which the definition of the shadow is clearest. If the distance is still further continually increased, the detail is lost, and finally only a vague shadow is perceptible, showing apparently the outlines of the hand.

This often-noted fact disagrees entirely with any theory of transverse vibrations, but can be easily explained on the assumption of material streams. When the hand is near and the velocity of the stream of particles very great, both bone and flesh are easily penetrated, and the effect due to the difference in the retardation of the particles passing through the heterogeneous parts can not be observed. The screen can fluoresce only up to a certain limited intensity, and the film can be affected only to a certain small degree. When the distance is increased, or, what is equivalent, when the intensity of the radiation is reduced, the more. resisting bones begin to throw the shadow first. Upon a further increase of the distance the flesh begins likewise to stop enough of the particles to leave a trace on the screen. But in all cases, at a certain distance, manifestly that which under the conditions of the experiment gives the greatest difference in the trajectories of the particles within *the range perceptible on the screen or film,* the clearest shadow is secured.

IV — the Rays Are All of One Kind

The preceding explains the apparent existence of rays of different kind; that is, of different rates of vibration, as it is asserted. In my opinion, the velocity and possibly the size of the particles both are different, and this fully accounts for the discordant results obtained in regard to the transparency of various bodies to these *rays.* I found, for example, in many cases that aluminum was less transparent than glass, and in some instances brass appeared to be very transparent as compared with other metallic bodies. Such observations showed that it was necessary, in making the comparison, to take rigorously equal thicknesses of the bodies and place them as closely together as possible. They also showed the fallacy of comparing results obtained with different bulbs.

V — Action on the Films

Many experiments with films of different thicknesses show that decidedly more detail is obtainable with a thick film than with a thin one. This appears to me to be a further evidence in support of the above views, as the result can be easily explained when considering the preceding remarks.

VI — the Behavior of Various Bodies in Reflecting the Rays

on which I have previously dwelt, will, if verified by other experimenters, leave no room for a doubt that the radiations are streams of some matter, or possibly of ether, as before observed.

VII — The Entire Absence of Refraction

and other features possessed by the light waves has, since Roentgen's announcement, not yet been satisfactorily explained. A trace at least of such an effect would be found if the rays were transverse vibrations.

VIII — The Discharge of Conductors

by the rays shows, in so far as l have been able to follow the researches of others, that the electrical charge is taken off by the bodily carriers. It is also found that the opacity plays an important part, and the observations are mostly in accord with the above views.

IX — The Source of the Rays

is, I find, always the place of the first impact of the cathodic stream, a second impact producing little or no rays. This fact would be difficult to account for unless streams of matter are assumed to exist.

X — Shadows in Space Outside of the Bulb

An almost crucial test of the existence of material streams is afforded by the formation of shadows in space at a distance from the bulb, to which I have called attention quite recently. I will presently refer to my preceding communication on this subject, and will only point out that such shadows could not be formed under the conditions described, except by streams of matter.

XI — All Bodies Are Transparent to Very Strong Rays

Experiments establish this fact beyond any doubt. With very intense radiations, I obtain, easily, impressions through what may be considered a great thickness of any metal. It is impossible to explain this on any theory of transverse vibrations. We can show how one or other body might allow the rays to pass through, but such explanations are not applicable to *all* bodies without exception. On the contrary, assuming material streams; such a result is unavoidable.

A great many other observations and facts might be added to the above, as further evidence in support of the above views. I have noted certain peculiarities of bodies obstructing a cathodic stream within the bulb. I have observed that the same rays are produced at all degrees of exhaustion and using bodies of vastly different physical properties, and have found a number of features in regard to the pressure, the vacuum, the residual gas, the material of the electrode, etc., all of which observations are more or less in accord with what I have stated before. I hope, however, that there is enough in the present lines to enlist the attention of others.

On the Hurtful Actions of Lenard and Roentgen Tubes

Electrical Review — May 5, 1897

The rapidly extending use of the Lenard and Roentgen tubes or Crookes bulbs as implements of the physician, or as instruments of research in laboratories, makes it desirable, particularly in view of the possibility of certain hurtful actions on the human tissues, to investigate the nature of these influences, to ascertain the conditions under which they are liable to occur and — what is most important for the practitioner — to render all injury impossible by the observance of certain rules and the employment of unfailing remedies.

As I have stated in a previous communication (see Electrical Review of December 2, 1896) no experimenter need be deterred from using freely the Roentgen rays for fear of a poisonous or deleterious action, and it is entirely wrong to give room to expressions of a kind such as may tend to impede the progress and create a prejudice against an already highly beneficial and still more promising discovery; but it can not be denied that it is equally uncommendable to ignore dangers now when we know that, under certain circumstances, they actually exist. I consider it the more necessary to be aware of these dangers, as I foresee the coming into general use of novel apparatus, capable of developing rays of incomparably greater power. In scientific laboratories the instruments are usually in the hands of persons skilled in their manipulation and capable of approximately estimating the magnitude of the effects, and the omission of necessary precautions is, in the present state of our knowledge, not so much to be apprehended; but the physicians, who are keenly appreciating the immense benefits derived from the proper application of the new principle, and th= numerous amateurs who are fascinated by the beauty of the novel manifestations, who are all passionately bent upon experimentation in the newly opened up fields, but many of whom are naturally not armed with the special knowledge of the electrician — all of these are much in need of reliable information from experts, and for these chiefly the following lines are written. However, in view of the still incomplete knowledge of these *rays,* I wish the statements which follow to be considered as devoid of authoritativeness, other than that which is based on the conscientiousness of my study and the faith in the precision of my senses and observations.

Ever since Professor Roentgen's discovery was made known I have carried on investigations in the directions indicated by him, and with perfected apparatus, producing rays of much greater intensity than it was possible to obtain with the usual appliances. Commonly, my bulbs were capable of showing the shadow of a hand on a phosphorescent screen at distances of 40 or 50 feet, or even more, and to the actions of these bulbs myself and several of my assistants were exposed for hours at a time, and although the exposures took place every day, not the faintest hurtful action was noted — as long as certain precautions were taken. On the contrary, be, it a coincidence, or an effect of the rays, or the result of some secondary cause present in the operation of the bulbs — as, for example, the generation of ozone — my own health, and that of two persons who were daily under the influence of the rays, more or less, has materially improved, and, whatever be the reason, it is a fact that a troublesome cough with which Z was constantly afflicted has entirely disappeared, a similar improvement being observed on another person.

In getting the photographic impressions or studying the rays with a phosphorescent screen, I employed a plate of thin aluminum sheet or a gauze of aluminum wires, which was interposed between the bulb and the person, and connected to the ground directly or through a condenser. I adopted this precaution because it was known to me, a long time before, that a certain irritation of the skin is caused by very strong streamers, which, mostly at small distance, are formed on the body of a person through . the electrostatic influence of a terminal of alternating high potential. I found that the occurrence of these streamers and their hurtful consequence was completely prevented by the employment of a conducting object, as a sheet of wire gauze placed and connected as described. It was observed, however, that the injurious effects mentioned did not seem to diminish gradually with the distance from the terminal, but ceased abruptly, and I could give no . other explanation for the irritation of the skin which would be as plausible as that which I have expressed; namely, that the effect was due to ozone, which was abundantly produced. The latter peculiarity mentioned was also in agreement with this view, since the generation of ozone ceases abruptly at a definite distance from the terminal, making it evident that a certain intensity of action is absolutely required, as in a process of electrolytic decomposition.

In carrying further my investigations, I gradually modified the apparatus in several ways, and immediately I had opportunities to observe hurtful influences following the exposures. Inquiring now what changes I had introduced, I found that I had made three departures from the plan originally followed; First, the aluminum screen was not used; second, a bulb was employed which, instead of aluminum, contained platinum, either as electrode or impact plate; and third, the distances at which the exposures took place were smaller than usual.

It did not require a long time to ascertain that the interposed aluminum sheet was a very effective remedy against injury, for a hand could be exposed for a long time behind it without the skin being reddened, which otherwise invariably and very quickly occurred. This fact impressed me with the conviction that, whatever the nature of the hurtful influences, it was in a large measure dependent either on an electrostatic action, or electrification, or secondary effects resulting therefrom, such as are attendant to the formation of streamers. This view afforded an explanation why an observer could watch a bulb for any length of time, as long as he was holding the hand in front of the body, as in examining with a fluorescent screen, with perfect immunity to all parts of hi& body, with the exception of the hand. It likewise explained why burns were produced in some instances on the opposite side of the body, adjacent to the photographic plate, whereas portions on the directly exposed part of the body, which were much nearer to the bulb, and consequently subjected to by far stronger rays, remained unaffected. It also made it easy to understand why the patient experienced a prickling sensation on the exposed part of the body whenever an injurious action took place. Finally, this view agreed with the numerous observations that the hurtful actions occurred when air was present, clothing, however thick, affording no protection, while they practically ceased when a layer of a fluid, quite easily penetrated by the rays, but excluding all contact of the air with the skin, was used as a preventive.

Following, now, the second line of investigation, I compared bulbs containing aluminum only with those in which platinum was used besides, ordinarily as impact body, and soon there were enough evidences on hand to dispel all doubt as to the latter metal being by far the more injurious. In support of this statement, one of the experiences may be cited which, at the same time, may illustrate the necessity — of taking proper precautions when operating bulbs of very high power. In order to carry out comparative tests, two tubes were constructed of an improved Lenard pattern, in size and most' other respects' alike. Both contained. a concave cathode or reflector of nearly two inches in diameter, and both were provided with an aluminum cap or window. In one of the tubes the cathodic focus was made to coincide with the center of the cap, in the other the cathodic stream was concentrated upon a platinum wire supported on a glass stem axially with the tube a little in front of the window, and in each case the metal of the latter was thinned down in the central portion to such an extent as to be barely able to withstand the inward air pressure. In studying the action of the tubes, I exposed one hand to that containing aluminum only, and' the other to the tube with the platinum wire. On turning on the former tube, I was surprised to observe that the aluminum window emitted a clear note, corresponding to the rhythmical impact of the cathodic stream. Placing the hand quite near the window, I felt distinctly that something warm was striking it. The sensation was unmistakable, and, quite apart. from the warmth felt, differed very much from that prickling feeling produced by streamers or minute sparks. Next I examined the tube with the platinum wire. No sound was emitted by the aluminum window, all the energy of the impact being seemingly spent on the platinum wire, which became incandescent, or else the matter composing the cathodic stream was so far disintegrated that the thin metal sheet offered no material obstruction to its passage. If big lumps are hurled against a wire netting with large meshes, there is considerable pressure exerted against the netting; if, on the contrary — for illustration — the lumps are very small as compared with the meshes, the pressure might not be manifest. But, although the window did not vibrate, I felt, nevertheless, again, and distinctly, that something was impinging against the hand, and the .sensation of warmth was stronger than in the previous case. In the action on the screen there was apparently no difference between the two tubes, both rendering it very bright, and the definition of the shadows was the same, as far as it was possible to judge. I had looked through the screen at the second tube a few times, only when something detracted my attention, and it was not until about 20 minutes later, when I observed that the hand exposed to it was much reddened and swollen. Thinking that it was due to some accidental injury, I turned again to the examination of the platinum tube; thrusting the same hand close to the window, and now I felt instantly a sensation of pain, which became more pronounced when the hand was placed repeatedly near the aluminum window. A peculiar feature was that the pain appeared to be seated, not at the surface, but deep in the tissues of the hand, or rather in the bones. Although the aggregate exposure was certainly not more than half a minute, I had to suffer severe pain for a few days afterward, and some time later I observed that all the *hair was* destroyed and that the nails on the injured hand had grown anew.

The bulb containing no platinum was now experimented with, more care being taken, but soon its comparative harmlessness was manifest, for, while it reddened the skin; the injury was not nearly as severe as with the other tube. The valuable experiences thus gained were: The evidence of something hot striking the exposed member; the pain *instantly* felt; the injury produced *immediately* after the exposure, and the increased violence due, in all probability, to the presence of the platinum.

Some time afterward I observed other remarkable actions at very small distances from powerful Lenard tubes. For instance, the hand being held near the window only for a few seconds, the skin seems to become tight, or else the muscles are stiffened, for some resistance is experienced in closing the fist, but upon opening and dosing it repeatedly the sensation disappears, apparently no ill effect remaining. I have, furthermore, observed a decided influence on the nasal discharge organs similar to the effects of a cold just contracted. But the most interesting observation in this respect is the following: When such a powerful bulb is watched for some time, the head of the observer being brought very close, he soon after that experiences a sensation so peculiar that no one will fail to notice it when once his attention is called to it, it being almost as positive as touch. If one imagines himself looking at something like a cartridge, for instance, in close and dangerous proximity, and just about to explode, he will get a good idea of the sensation produced, only, in the case of the cartridge, one can not render himself an account where the feeling exactly resides, for it seems to extend all over the body, this indicating that it comes from a general awareness of danger resulting from previous and manifold experiences, and not from the anticipation of an unpleasant impression directly upon one of the organs, as the eye or the ear; but, in the case of the Lenard bulb, one can at once, and with precision, locate the sensation; it is in the head. Now, this observation might not be of any value except, perhaps, in view of the peculiarity and acuteness of the feeling, were it not that exactly the same sensation is .produced when working for some time with a noisy spark gap, or, in general, when exposing the ear to sharp noises or explosions. Since it seems impossible to imagine how the latter could cause such a sensation in any other way except by directly impressing the organs of hearing, I conclude, that a Roentgen or Lenard tube, working in perfect silence as it may, nevertheless produces violent explosions or reports and concussions, which; though they are inaudible, take some material effect upon the bony structure of the head. Their inaudibility may be sufficiently explained by the well founded assumption that not the air, but some finer medium, is concerned in their propagation.

But it was in following up the third line of inquiry into the nature of these hurtful actions, namely, in studying the influence of distance, that the most important fact was unearthed. To illustrate it popularly, I will say that the Roentgen tube acts exactly like a source of intense heat. If one places the hand near to a red-hot stove, he may be instantly injured. If he keeps the hand at a. certain small distance, he may be able to withstand the rays for a few minutes or more, and may still be injured by prolonged exposure; but if he recedes only a little farther, where the heat is slightly less, he may withstand the heat in comfort and any length of time without receiving any injury, the radiations at that distance being too weak to seriously interfere with the life process of the skin. This is absolutely the way such a bulb acts. Beyond a certain distance no hurtful effect whatever is produced on the skin, no matter *how long* the exposure. The character of the burns is also such as might be expected from a source of high heat. I have maintained, in all deference to the opinions of others, that those who have likened the effect`s on the skin arid tissues to sunburns have misinterpreted them. There is no similarity in this respect, except in so far 'as the reddening and peeling of the skin is concerned; which may result from innumerable causes. The burns, when slight, rather resemble those people often receive when working .close to a strong fire. But when the injury is severe, it is in all appearances like that received from contact with fire or from a red-hot iron. There may be no period of incubation at all, as is evident from the foregoing, remarks; the rays taking effect immediately, not to say instantly. In a severe case the skin gets deeply colored and blackened in places, and ugly; ill-foreboding blisters form; thick layers come off, exposing the raw flesh, which; for a time; discharges freely, Burning pain, feverishness and such symptoms are of course but natural accompaniments. One single injury of this kind, in the abdominal region, to a dear and zealous assistant — the only accident that ever happened to any one but myself in all my laboratory experience — I had the misfortune to witness. It occurred before all these and other experiences were gained; following directly an exposure of five. minutes at. the fairly safe distance of 11 inches to a very highly-charged platinum tube, the protecting aluminum screen having been unfortunately omitted, and it was such as to fill me with the gravest apprehensions. Fortunately, frequent warm baths, free application of vaseline, cleaning and general bodily care soon repaired the ravages of the destructive agent, and I breathed again freely. Had I known more than I did of these injurious actions, such unfortunate exposure would not have been made; had I known less than I did, it might have been made at a smaller distance, and a serious, perhaps .irremediable, injury might have resulted. ,

I am using the first opportunity to comply with the bitter duty of recording the accident. I hope that others will do likewise, so that the most complete knowledge of these dangerous. actions may soon be acquired. My apprehensions led me to consider, with keener interest than I would have felt otherwise, what the probabilities were in such. a case of the internal tissues being seriously injured. I came to the very comforting conclusion that, no matter what the rays are ultimately recognized to be, practically all their destructive energy must spend itself on the surface of the body, the internal tissues being, in all probability, safe, unless the bulb would be placed in very close proximity to the skin, or else, that rays of far greater intensity than now producible were generated. There are many reasons why this should be so, some of which will appear dear from my foregoing statements referring to the nature of the hurtful agencies, but I may be able to cite new facts in support of this view. A significant feature of the case reported may be mentioned. It was observed that on three places, which were covered by thick bone buttons, the skin was entirely unaffected, while it was entirely destroyed under each of the small holes in the buttons. Now, it was impossible for the rays, as investigation showed, to reach these points of the skin in straight lines drawn from the bulb, and this would seem to indicate that not all the injury was due to the *rays* or radiations under consideration, which unmistakably propagate in straight lines, but that, at least in part, concomitant causes were responsible. A further experimental demonstration of this fact may be obtained in the following manner: The experimenter may excite a bulb to a suitable and rather small degree, so as to illuminate the fluorescent screen to a certain intensity at a distance of, say, seven inches. He may expose his hand at that distance, and the skin will be reddened after a certain duration of exposure. He may now force the bulb up to a much higher power, until, at a distance of 14 inches, the screen is illuminated even stronger than it was before at half that distance: The rays are now evidently stronger at the greater distance, and yet he may expose the hand a very long time, and it is safe to assert that he will not be injured. Of course, it is possible to bring forth arguments which might deprive the above demonstration of force. So, it might. be stated, that the actions on the screen or photographic plate do not give us an idea as to the density and other quantitative features of the rays, these actions being entirely of a qualitative character. Suppose the rays are formed by streams of material particles, as I believe, it is thinkable that it might be of no particular consequence, in so far as the visible impression on the screen or film is concerned, whether a trillion of particles per square millimeter strike the sensitive layer or only **a** million, for example; but with the actions on the skin it is different; theca must surely and very materially depend on the quantity of the streams.

As soon as the before-mentioned fact was recognized, namely, that beyond a certain distance even the most powerful tubes are incapable of producing injurious action; no matter how long the exposure may last, it became important to ascertain the safe distance. Going over all my previous experiences, I found that, very frequently, I have had tubes which at a distance of 12 feet, .for illustration, gave a strong impression of the chest of a person with an exposure of a few minutes, and many times persons have been subjected to the rays from these tubes at a distance of from

18 to 24 inches, the time of exposure varying from 10 to 45 minutes, and never the faintest trace of an injurious action was observed. With such tubes I have even made long exposures at distances of 14 inches, always, of course, through a thin sheet or wire gauze of aluminum connected to the ground, and, in each case, observing the precaution that the metal would not give any spark when the person was touching it with the hand, as it might sometimes be when the electrical vibration is of extremely high frequence, in, which we a ground connection, through a condenser of proper capacity, should be resorted to. In all these instances bulbs containing only aluminum were used; and I therefore still lack sufficient data to form an enact idea of what distance would have been safe with a platinum tube. Froth the case previously cited, we see that a grave injury, resulted at a distance of 11 inches, but I believe that, had the protecting screen been used, the injury, if any, would have-been very slight. Taking all my experiences together, I am convinced that no serious injury can result if the distance is greater than 16 inches and the impression is taken in the manner I have described.

Having been successful in a number of lines of inquiry pertaining to this new department of science, I am able at present to form a broader view of the actions of the bulbs, which; I hope, will soon assume a quite definite shape. For the present, the following brief statement may be sufficient. According to the evidences I am obtaining, the bulb, when in action, is emitting a stream of small material particles. There are some experiments which seem to indicate that these particles start from the outer wall of the bulb; there are others which seem to prove that there is an actual penetration of the wall, and, in the case of a thin aluminum window, I have now not the least doubt that some of the finely disintegrated cathodic matter is actually forced through. These streams may simply be projected to a great distance, the velocity gradually diminishing without the formation of any waves, or they may give rise to concussions and longitudinal waves. This, for the present consideration, is entirely immaterial, but, assuming the existence of such streams of particles, and disregarding such actions as might be due to the properties, chemical or physical, of the projected matter, we have to consider the following specific actions:

First. There is the thermal effect. The temperature of the electrode or impact body does not in any way give us an idea of the degree of heat of the particles, but, if we consider the probable velocities only, they correspond to temperatures which may be as high as 100,000 degrees centigrade. It may be sufficient that the particles are simply at a high temperature to produce an injurious action, and, in fact, many evidences point in this direction. But against this is the experimental fact that we can not demonstrate such a transference of heat, and no satisfactory explanation is found yet, although, in carrying my investigations in this direction, I have arrived at some results.

Second, there is the purely electrical effect. We have absolute experimental evidence that the particles or rays, to express myself generally, convey an immense amount of electricity, and I have even found a way of how to estimate and measure that amount. Now it is likewise possible that the mere fact of these particles being highly electrified is sufficient to cause the destruction of the tissue. Certainly, on contact with the skin, the electrical charges will be given off, and may give rise to strong and destructive local currents in minute paths of the tissue. Experimental results are in accord with this view, and, in pushing my inquiry in this direction, I have been still more successful than in the first. Yet, while as I have suggested before, this view explains best the action on a sensitive layer, experiment shows that, when the supposed particles traverse a grounded plate, they are not deprived entirely of their electrification, which is not satisfactorily explained.

The *third* effect to be considered is the electro-chemical. The charged particles give rise to an abundant generation of ozone and other gases, and these we know, by experiment, destroy even such a thing as rubber, and are, therefore, the most likely agent in the destruction of the skin, and the evidences are strongest in this direction, since a small layer of a fluid, preventing the contact of gaseous matter with the skin, seems to stop all action.

The *last* effect to be considered is the purely mechanical. It is thinkable that material particles, moving with great speed, may, merely by a mechanical impact and unavoidable heating at such speeds, be sufficient to deteriorate the tissues, and in such a case deeper layers might also be injured, whereas it is very probable that no such. thing would occur if any, of the former explanations would be found to hold.

Summing up my experimental experiences. and the conclusions derived from them, it would seem advisable,. first, to abandon, the use of bulbs containing platinum; second, to substitute for them a properly constructed Lenard tube, containing pure aluminum only, a tube of this kind having, besides, the advantage that it can be constructed with great mechanical precision, and therefore is capable of producing .much sharper impressions; third, to use a protecting screen of aluminum sheet, as suggested, or, instead of this, a wet cloth or a layer of a fluid; fourth, to make the exposures at distances of, at least, 14 inches, and preferably expose longer at a larger distance.

On the Source of Roentgen Rays and the Practical Construction and Safe Operation of Lenard Tubes

Electrical Review — August 11, 1897

I have for some time felt that a few indications in regard to the practical construction of Lenard tubes of improved designs, a great number of which I have recently exhibited before the New York Academy of Sciences (April 6, 1897), would be useful and timely, particularly as by their proper construction and use much of the danger attending the experimentation with the rays may be avoided. The simple precautions which I have suggested in my previous communications are seemingly disregarded, and cases of injury to patients are being almost daily reported, and in view of this only, were it for no other reason, the following lines, referring to this subject, would have been written before had not again pressing and unavoidable duties prevented me from doing so. A short and, I may say, most unwelcome interruption of the work which has been claiming my attention makes this now possible. However, as these opportunities are scarce, I will utilize the present to dwell in a few words on some other matters in connection with this subject, and particularly on a result of importance which I have reached some time ago by the aid of such a Lenard tube, and which, if I am correctly informed, I can. only in part consider as my own, since it seems that practically it has been expressed in other words by Professor Roentgen in a recent communication to the Academy of Sciences of Berlin. The result alluded to has reference to the much disputed question of the source of the Roentgen rays. As will be remembered, in the first announcement of his discovery, Roentgen was of the opinion that the rays which affected the sensitive layer emanated from the fluorescent spot on the glass wall of the bulb; other scientific men next made the cathode responsible; still others the anode, while some thought that the rays were emitted solely from fluorescent powders of surfaces, and speculations, mostly unfounded, increased to such an extent that, despairingly, one , would exclaim with the poet:

"O glucklich wer noch hoffen kann,
Aus diesem Meer des Irrtums aufzutauchen!"

My own experiments led me to recognize that, regardless of the location, the chief source of, these rays was the place of the *first* impact of .the projected stream of particles within the bulb. This was merely a broad statement, of which that of Professor Roentgen was a special case, as in his first experiments the fluorescent spot on the glass wall was, incidentally, the place of the first impact of the cathodic stream. Investigations carried on up to the present day have only confirmed the correctness of the above opinion, and the place of the first collision of the stream of particles — be it an anode or independent impact body, the glass wall or an aluminum window — is still found to be: the principal source of the rays. But, as will be seen presently, it is not the only source.

Since recording the above fact my efforts were directed to finding answers to the following questions: First, is it necessary that the impact body should be within the tube? Second, is it required that the obstacle in the path of the cathodic stream should be a solid or liquid? And, third, to what extent is the velocity of the stream necessary for the generation of and influence upon the character of the rays emitted?

In order to ascertain whether a body located outside of the tube and in the path or in the direction of the stream of particles was capable of producing the same peculiar phenomena as an object located inside; it appeared necessary to first show that there is an actual penetration of the particles through the wall, or otherwise that' the actions of the supposed streams; of whatever nature they might, be, were sufficiently pronounced in the outer region close to the wall of the bulb as to produce some of the effects which are peculiar to a cathodic stream. It was not difficult to obtain with a properly prepared Lenard tube, having an exceedingly thin window, many and at first surprising evidences of this character. Some of these have already, been pointed out; and it is thought sufficient to cite here one more which I have since observed. In the hollow aluminum cap A of a tube as shown in diagram Fig. 1, which will be described in detail, I placed a half-dollar silver piece, supporting it at a small distance from and parallel to the window or bottom of the cap by strips of mica in such a' manner . that it was not touching the metal of the tube, an air space being left all around it: Upon exciting the bulb for about 30 to 45 seconds by the secondary discharge of a powerful coil of a novel type now well known, it was found that the silver piece was rendered so hot as to actually scorch the hand; yet the aluminum window, which offered a very insignificant obstacle to the cathodic stream, was only moderately warmed. Thus it was shown that the silver alloy, owing to its density and thickness, took up most of the energy of the impact, being acted upon by the particles almost identically as if it had been inside of the bulb, and, what is more, indications were obtained, by observing the shadows, that it behaved like a second source of the rays, inasmuch as the outlines of the shadows, instead of being sharp and clear as when the half-dollar piece was removed, were dimmed. It was immaterial for the chief object of the inquiry to decide by more exact methods whether the cathodic particles actually penetrated the window, or whether a new and separate stream was projected from the outer side of the window. In my mind there exists not the least doubt that the former was the case, as in this respect I have been able to obtain numerous additional proofs, upon which I may dwell in the near future.

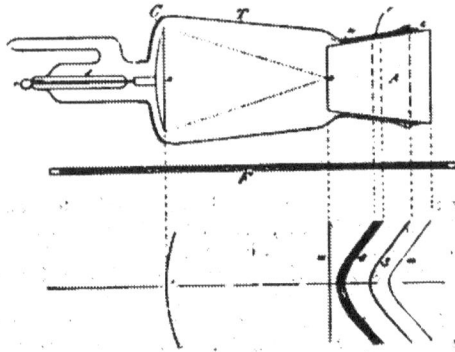

Fig. 1. — Illustrating an Experiment Revealing the Real Source of the Roentgen Rays.

I next endeavored to ascertain whether it was necessary that the obstacle outside was, as in this case, a solid body, or a liquid, or broadly, a body of measurable dimensions, and it was in investigating in this direction that I came upon the important result to which I referred in the introductory statements of this communication. I namely observed rather accidentally, although I was following up a systematic inquiry, what is illustrated in

.diagram Fig. 1. The diagram shows a Lenard tube of improved design, consisting of a tube T of thick glass tapering towards the open end, or neck n, into which is fitted an aluminum cap A, and a spherical cathode $e,$ supported on a glass stem $s,$ and platinum wire w sealed in the opposite end of the tube as usual. The aluminum cap A, as will be observed, is not in actual contact. with the ground-glass wall, being held at a small distance from the latter by a narrow and continuous ring of tinfoil r. The outer space between the glass and the cap A is filled with cement $c,$ in a manner which I shall later describe. F is a Roentgen screen such as is ordinarily used in making the observations.

Now, in looking upon the screen in the direction from F to T, the dark lines indicated on the lower part of the diagram were seen on the illuminated background. The curved line e and the straight line W were, of course, at once recognized as the outlines of the cathode a and the bottom of the cap A respectively, although, in consequence of a confusing optical illusion, they appeared mush closer together than they actually were. For instance, if the distance between a and o was five inches, these lines would appear on the screen about two inches apart, as nearly as I could judge by the eye. This illusion may be easily explained and is quite unimportant, except that it might be of some moment to physicians to keep this fact in mind when making examinations with the screen as, owing to the above effect, which is sometimes exaggerated to a degree hard to believe, a completely erroneous idea of the distance of the various parts of the object under 'examination might be gained, to the detriment of the surgical operation. But while the lines a and W were easily accounted for, the curved lines $t, g,$ a were at first puzzling. Soon, however, it was ascertained that the faint line a was the shadow of the edge of the aluminum cap, the much darker line g that of the rim of the glass tube T, and t the shadow of the tinfoil ring r. These shadows on the screen F clearly showed that the agency which affected the fluorescent material was proceeding from the space outside of the bulb towards the' aluminum cap, and chiefly from the region through which the primary disturbances or streams emitted from the tube through the window were passing, which observation could not be explained in a more plausible manner than by assuming that the air and dust particles outside, in the path of the projected streams, afforded an obstacle to their passage and gave rise to impacts and collisions spreading through the air in all directions, thus producing continuously new sources of the rays. It is this fact which; in his recent communication before mentioned, Roentgen has brought out. So, at least, I have interpreted his reported statement that the rays emanate from the irradiated air. It now remains to be shown whether the air, from which carefully all foreign particles are removed, is capable of behaving as an impact body and source of the rays, in order to decide whether the generation of the latter is dependent on the presence in the air of impact particles of measurable dimensions. I have reasons to think so.

With the knowledge of this fact we are now able to form a more general idea of the process of generation of the radiations which have been discovered by Lenard and Roentgen. It may be comprised in the statement that the streams of minute material particles projected from an electrode with great velocity in encountering obstacles wherever they may be, within the bulb, in the air or other medium or in the sensitive layers themselves, give rise to rays or radiations possessing many of the properties of those known as light. If this physical process of generation of these rays is undoubtedly demonstrated as true, it will have most important consequences, as it will induce physicists to again critically examine many phenomena which are presently attributed to transverse ether waves, which may lead to a radical modification of existing views and theories in regard to these phenomena, if not as to their essence so, at least, as to the mode of their production. '

My effort to arrive at an answer to the third of the above questions led me to the establishment, by actual photographs, of the dose relationship which exists between the Lenard and Roentgen rays. The photographs bearing on this point were exhibited of a meeting of the New York Academy of Sciences — before referred to — April 6, 1897, but, unfortunately, owing to the shortness of my address, arid concentration of thought on other matters, I omitted what was most important; namely, to describe the manner in which these, photographs were obtained, an oversight which I was able to only partially repair the day following. I did, however, on that occasion illustrate and describe experiments in which was shown the deflectibility of the Roentgen rays by a magnet, which establishes a still closer relationship, if not identity, of the rays named after these two discoverers. But the description of these experiments in detail, as well as of other investigations and results in harmony with and restricted to the subject I brought before that scientific body, will appear in a longer communication which I am slowly preparing.

To bring out clearly the significance of the photographs in question, I would recall that, in some of my previous contributions to scientific societies, I have endeavored to dispel a popular opinion before existing that the phenomena known as those of Crookes were dependent on and indicative of high vacua. With this object in view, I showed that phosphorescence and most of the phenomena in Crookes bulbs were producible at greater pressures of the gases in the bulbs by the use of much higher or more sudden electro-motive impulses. Having this well demonstrated fact before me, I prepared a tube in the manner described by Lenard in his first classical communication on this subject. The tube was exhausted to a moderate degree, either by chance or of necessity, and it was found that, when operated by an ordinary high-tension coil of a low rate of change in the current, no rays of any of the two kinds could be detected, even when the tube was so highly strained .as to become very hot in a few moments. Now, I expected that, if the suddenness of the impulses through the bulb were sufficiently increased, rays would be emitted. To test this I employed a coil of a type which I have repeatedly described, in which the primary is operated by the discharges of a condenser. With such an instrument any desired suddenness of the impulses may be secured, there being practically no limit in this respect, as the energy accumulated in the condenser is the most violently explosive agent we know, and any potential or electrical pressure is obtainable: Indeed, I found that in increasing the suddenness of the electro-motive, impulses through the tube — without, however, increasing, but rather diminishing the total energy conveyed to it — phosphorescence was observed and rays began to appear, first the feebler Lenard rays and later, by pushing the suddenness far enough, Roentgen rays of great intensity, which enabled me to obtain photographs showing the finest texture of the bones. Still, the same tube, when again operated with the ordinary coil of a low rate of change in the primary current, emitted practically no rays, even when, as before stated, much more energy, as judged from the heating, was passed through it. This experience, together with the fact that I have succeeded in producing. by the use of immense electrical pressures, obtainable with certain apparatus designed for this express purpose, some impressions in free air, have led me to the conclusion that in lightning discharges Lenard and Roentgen *rays* must be generated at ordinary, atmospheric pressure.

At this juncture I realize, by a perusal of the preceding lines, that my scientific interest has dominated the practical, and that the following remarks must be devoted to the primary object of this communication — that is, to giving some data for the construction to those engaged in the manufacture of the tubes and, perhaps, a few useful hints to practicing physicians who are dependent on such information. The foregoing was, nevertheless, not lost for this object, inasmuch as it has shown how much the result obtained depends on the proper construction of the instruments, for, with ordinary implements, most of the above observations could not have been made.

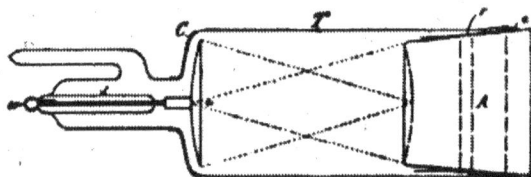

Fig. 2. — Improved Lenard Tube.

I have already described the form of tube illustrated in Fig. 1, and in Fig. 2 another still further improved design is shown. In this case the aluminum cap A, instead of having a straight bottom as before, is shaped spherically, the renter of the sphere coinciding with that of the electrode *e*, which itself, as in Fig. 1, has it's focus in the center of the window of cap A, as indicated by the dotted lines. The aluminum cap A has a tinfoil ring *r*, as that in Fig. 1, or else the metal of the cap is spun out on that place .so as to afford a bearing of small surface between the metal and the glass. This is an important practical detail as, by making the bearing surface small, the pressure per unit of area is increased and a more perfect joint made. The ring *r* should be first spun out and then ground to fit the neck of the bulb. If a tinfoil ring is used instead, it may be cut out of one of the ordinary tinfoil caps obtainable in the market, care being taken that the ring is very smooth.

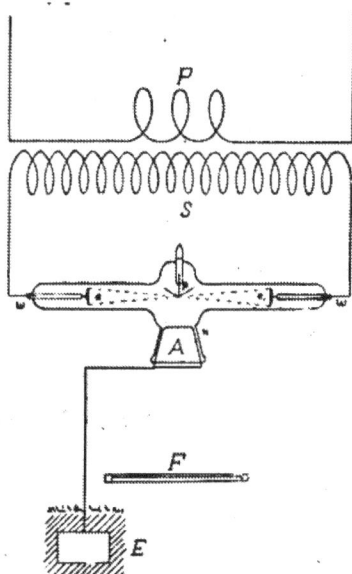

Fig. 3. — Illustrating Arrangement
with Improved Double-Focus Tube
for Reducing the Injurious Actions.

In Fig. 3 I have shown a modified design of tube which, as the two types before described, was comprised in the collection I exhibited. This, as will be observed, is a double-focus tube, with impact plates of iridium alloy and an aluminum cap A opposite the same. The tube is not shown because of any originality in design, but simply to illustrate a practical feature. It will be noted that the aluminum caps in the tubes described are fitted inside of the necks and not outside, as is frequently done. Long experience has demonstrated that it is practically impossible to maintain a high vacuum in a tube with an outside cap. The only way I have been able to do this in a fair measure is by cooling the cap by a jet of air, for instance, and observing the following precautions: The air jet is first turned on slightly and upon this the tube is excited. The current through the latter, and also the air pressure, are then gradually increased and brought to the normal working condition. Upon completing the experiment the air pressure and current through the tube are both gradually reduced and both so manipulated that no great differences in temperature result between the glass and aluminum cap. If those precautions are not observed the vacuum will be immediately impaired in consequence of the uneven expansion of the glass and metal.

With tubes, as these presently described, it is quite unnecessary to observe this precaution if proper care is taken in their preparation. In inserting the cap the latter is cooled down as low as it is deemed advisable without endangering the glass, and it is then, gently pushed in the neck of the tube, taking care that it sets straight.

The two most important operations in the manufacture of such a tube are, however, the thinning down of the aluminum window and the sealing in of the cap. The metal of the latter may be one thirty-second or even one-sixteenth of an inch thick, and in such case the central portion may be thinned down by a countersink tool about one-fourth of an inch in diameter as far as it is possible without tearing the sheet. The further thinning down may then be done by hand with a scraping tool; and, finally, the metal should be gently beaten down so as to surely close the pores which might permit a slow leak. Instead of proceeding in this way I have employed a cap with a hole in the center, which I have closed with a sheet of pure aluminum a few thousandths of an inch thick, riveted to the cap by means of a washer of thick metal, but the results were not quite as satisfactory.

In sealing the cap I have adopted the following procedure: The tube is fastened on the pump in the proper position and exhausted until a permanent condition is reached. The degree of exhaustion is a measure of perfection of the joint. The leak is usually considerable, but this is not so serious a defect as might be thought. Heat is now gradually applied to the tube by means of a gas stove until a temperature up to about the boiling point of sealing wax is reached. The space between the cap and the glass is then filled with sealing wax of good quality; and, when the latter begins to boil, the temperature is reduced to allow its settling in the cavity. The heat is then again, increased, and this process of heating and cooling is repeated several times until the entire cavity, upon reduction of the temperature, is found to be filled uniformly with the wax, all bubbles having disappeared. A little more wax is then put on the top and the exhaustion carried on for an hour or so, according to the capacity of the pump, by application of moderate heat much below the melting point of the wax.

A tube prepared in this manner will maintain the vacuum very well, and will last. indefinitely. If not used for a few months, it may gradually lose the high vacuum, but it can be quickly worked up. However, if after long use it becomes necessary to clean the tube, this is easily done by gently warming it and taking off the cap: The cleaning may be done first with acid, then with highly diluted alkali, next with distilled water, and finally with pure rectified alcohol.

These tubes, when properly prepared, give impressions much sharper and reveal much more detail. than those of ordinary make. It is important for the clearness of the impressions that the electrode should be properly shaped, and that the focus should be exactly in the center of the cap or slightly inside. In fitting in the cap, the distance from the electrode should be measured as exactly as possible. It should also be remarked that the thinner the window, the sharper are the impressions, but it is not advisable to make it too thin, as it is apt to melt in a point on turning on the current.

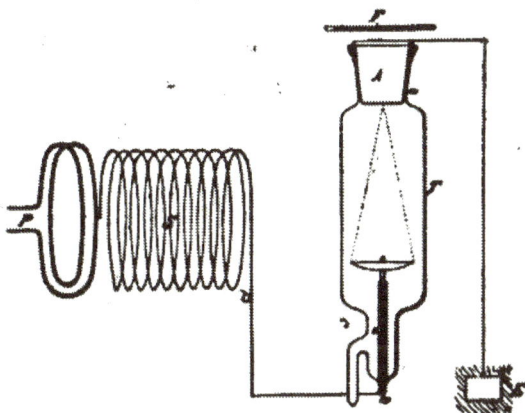

Fig. 4. — Illustrating Arrangement with a Lenard Tube for Safe Working at Close Range.

The above advantages are not the only ones which these tubes offer. They are also better adapted for purposes of examination by surgeons, particularly if used in the peculiar manner illustrated in diagrams Fig. 3 and Fig. 4, which are self-explanatory. It will be seen that in each of these the cap is connected to the ground. This decidedly diminishes the injurious action and enables also to take impressions with very short exposures of a few seconds only at dose range, inasmuch as, during the operation of the bulb, one can easily touch the cap without any inconvenience, owing to the ground connection. The arrangement shown in Fig. 4 is particularly advantageous with a form of single terminal, which coil I have described on other occasions and which is diagrammatically illustrated, P being the primary and S the secondary. In this instance the high-potential terminal is. connected to the electrode, while the cap is grounded. The tube may be placed in the position indicated in the drawing, under the operating table and quite close or even in contact with the body of the patient, if the impression requires only a few seconds as, for instance, in examining parts of the members. I have taken many impressions with such tubes and have observed no injurious action, but I would advise not to expose for longer than two or three minutes at very short distances. In this respect the experimenter should bear in mind what I have stated in previous communications. At all events it is certain that, in proceeding in the manner described, additional safety is obtained and the process of taking impressions much quickened. To cool the cap, a jet of air may be used, as before stated, or else a small quantity of water may be ,poured in the cap each time when an impression is taken. The water only slightly impairs the action of the tube, while it maintains the window at a safe temperature. I may add that the tubes are improved by providing back of the electrode a metallic coating C, shown in Fig. 3 and Fig. 4.

www.ingramcontent.com/pod-product-compliance
Lightning Source LLC
Chambersburg PA
CBHW021912040426
42447CB00007B/821